SIXTH EDITION

The
Physical Education
Curriculum

Jim L. Stillwell
Arkansas State University

Carl E. Willgoose

CBS Publishers & Distributors Pvt. Ltd.

New Delhi • Bengaluru • Chennai • Kochi • Mumbai • Pune
Hyderabad • Kolkata • Nagpur • Patna • Vijayawada

WAVELAND
PRESS, INC.
Long Grove, Illinois

The Physical Education Curriculum

Waveland ISBN: 978-1-57766-388-8

CBS Reprint: 2015

CBS ISBN: 978-81-239-2664-3

Published by:
Satish Kumar Jain for CBS Publishers & Distributors Pvt. Ltd.,
4819/XI Prahlad Street, 24 Ansari Road, Daryaganj, New Delhi - 110002
delhi@cbspd.com, cbspubs@airtelmail.in • www.cbspd.com
Ph.: 23289259, 23266861, 23266867 • Fax: 011-23243014

Corporate Office: 204 FIE, Industrial Area, Patparganj, Delhi - 110 092
Ph: 49344934 • Fax: 011-49344935
E-mail: publishing@cbspd.com • publicity@cbspd.com

Branches:
• *Bengaluru:* 2975, 17th Cross, K.R. Road, Bansankari 2nd Stage, Bengaluru - 70 Ph: +91-80-26771678/79 • Fax: +91-80-26771680
  E-mail: cbsbng@gmail.com, bangalore@cbspd.com
• *Chennai:* No. 7, Subbaraya Street, Shenoy Nagar, Chennai - 600030
  Ph: +91-44-26681266, 26680620 • Fax: +91-44-42032115
  E-mail: chennai@cbspd.com
• *Kochi:* 36/14, Kalluvilakam, Lissie Hospital Road, Kochi - 682018
  Ph: +91-484-4059061-65 • Fax: +91-484-4059065
  E-mail: cochin@cbspd.com
• *Mumbai:* 83-C, Dr. E. Moses Road, Worli, Mumbai - 400018
  Ph: +91-9833017933, 022-24902340/41 • E-mail: mumbai@cbspd.com
• *Pune:* Bhuruk Prestige, Sr. No. 52/12/2+1+3/2,
  Narhe, Haveli (Near Katraj-Dehu Road Bypass), Pune - 411041
  Ph: +91-20-64704058/59, 32342277 • E-mail: pune@cbspd.com

Representatives:
• Hyderabad: 0-9885175004   • Kolkata: 0-9831437309, 0-9051152362
• Nagpur: 0-9021734563   • Patna: 0-9334159340
• Vijayawada: 0-9000660880

Printed at: India Binding House, Noida, U.P.

CONTENTS

PREFACE

The purpose of this book is to help awaken readers to the genuine need for physical education in today's world and to assist them in the process of developing a curriculum for grades K–12. The primary thrust is directed toward curriculum improvement. We firmly believe that implementing sound curriculum development practices will contribute to quality programs that meet students' need for physical activity as well as the public's demand for educational accountability.

Carl E. Willgoose began the preface of the first edition of this text by quoting Marcus Aurelius, second century emperor and philosopher, who said that one should always observe that everything is the result of change and get used to thinking that there is nothing nature loves so well as to change existing forms and to make new ones like them. This statement could have been made today by anyone involved in curriculum planning at any level, since it embodies the essential truth that underlies curriculum development in physical education. Societies change, as do the people within these societies. Schools change, as do the students within these schools. As a result, to better meet the ever-changing needs and interests of students, physical education curricula also must change.

Not surprisingly, with this edition comes change. The linear organization has not changed because it provides an easy-to-follow format that heightens the reader's understanding of (1) what a curriculum is and (2) how a curriculum should be developed, implemented, and evaluated. In addition to a complete updating of content, including photos and illustrations, specific changes include:

- A broad discussion of American society
- A clearly defined presentation of general, specific, and behavioral objectives
- A more detailed presentation of the ten most common physical education curriculum models
- The deletion of chapter 7 and the incorporation of a more practical discussion of curriculum guides in chapter 6

- A revised and shortened chapter on children with disabilities
- Expanded content in chapters 7 and 8, with activity examples in each category
- A presentation of topical issues in interscholastic athletic programs
- A broader discussion, with examples, of student, teacher, and program evaluation in both elementary and secondary settings

Developing and implementing a sound physical education curriculum present a challenge to each educator. The public concern for educational accountability is reason enough to strive to meet this challenge. Moreover, to do less is to risk missing the opportunity to help students discover Dubos' concept of the celebrated life—a life of direction, adaptation, excitement, and human awareness filled with spirited individuals who appreciate the complex relationship between being able and being well.

Acknowledgments

As is the case with any text, many individuals who contributed to the development and completion of this volume need to be recognized. First, a professional thanks goes to Dr. Carl E. Willgoose for having written the first four editions. From his solid foundation, this updated sixth edition was made possible.

To the many students and colleagues who brought me to a stage and place in my professional career where I felt comfortable sharing my beliefs about what it is I do, I extend a sincere thank you.

Finally, I offer a very special thanks to my wife, Nita, for her understanding, acceptance, and support during my hours at the computer.

1

SOCIETY AND THE CELEBRATION OF LIFE

Outline

Outcomes

After reading and studying this chapter, you should be able to:

- Define
  The celebrated life
  Multicultural education
  Obesity
  Overweight
  Value illness
  Well-being
  Wellness
- Provide a sound philosophy of physical education.
- Identify changes that affect our society, including cultural diversity, longevity, and technology, as well as the implications arising from these changes.
- Develop a strategy to combat value illness.
- Justify the need for living a healthy lifestyle.
- Describe the wellness approach to living.

3

From the classical Indian Sanskrit, there is an enlightening expression which says that a day well lived is what determines all tomorrows. This concept has been around a long time. As applied to our subject matter, optimum health and human functioning have always depended on the routines of one's daily life. Dubos (1981) wrote enthusiastically about *the celebrated life*, a life of direction, adaptation, excitement, and human awareness; a life of spirited men and women who acknowledge their fragilities and sensitivities and appreciate, beyond a doubt, the complex relationship between being able and being well.

Unfortunately, society has not always defined well-being in terms of human vitality and productivity. Historically, well-being has been defined as freedom from disease, discussed in terms of infectious organisms, degenerative conditions, and defective organs. In recent years, disease has been more broadly considered as an organism's total lack of ease, a dis-ease with numerous behavioral overtones of a psychosomatic nature, a dis-ease relating to such significant human movement deficiencies as chronic fatigue, lower back pain, hypertension, and obesity. This shortcoming in functional ability and the physical means to perform has been a major deterrent to the advancement of our civilization.

Those who embrace this celebration of life concept can be reasonably optimistic about supporting the objective of a fully awakened, enlightened, and able lifestyle. Such individuals also have a beyond-the-self contribution to make in advancing human welfare.

OUR FREE SOCIETY

The ultimate health of Americans and the inhabitants of the rest of the world depends on informed individuals who understand both the benefits and responsibilities of living in a free society—individuals preparing to join in seeking new and better solutions to the age-old problems confronting humanity. In terms of educational programs, there are problems pertaining to where the emphasis is placed. Has civilization become so organized, with an overspecialized development of the intellect, that it has become separated from the senses and will soon be incapable of continued functioning? Has the intellectual pursuit for truth been carried to such an extreme that it overshadows our cultural and physical aspirations, which are so much a part of a rich and full life?

When the American Alliance for Health, Physical Education, Recreation, and Dance (AAHPERD) joined with the governing boards of 23 professional education associations to express a sense of direction and renewed commitment to a more complete and balanced education for all, two objectives of significance to physical education were clearly highlighted:

- To express oneself through the arts and understand the artistic expression of others
- To apply knowledge about health, nutrition, and physical activity

These objectives speak to an awareness of the human potential, whereby men and women must step back and look at themselves in the total scheme of things. In doing so, the pattern of life, as influenced by physical education, takes on considerable meaning. Civilization can easily stagnate due to the cumulative effect of the forces around it, especially if individuals rigidly adhere to established routines. This is particularly true with regard to the practices of today's leaders in the many specialized fields of education. The task of formulating a new culture, with new goals, sanctions, patterns, and responsibilities, is an issue for these leaders.

Consider the leaders of the Roman Empire, who had a standard of living unparalleled in the history of the world. However, they did not choose to think beyond their materialistic desires. Because Romans were too resistant to change, the Roman culture slowly dissolved. Soft in mind, spirit, and flesh, they became viewers of the passing scene rather than participants in its evolution. The Romans chose to be smug, complacent, and satisfied at a time when the world demanded sensitivity, change, and struggle.

The true worth of physical education, or any education for that matter, is determined by how it affects the values, judgments, and commitments of the individual members of the society in which it is taught. In terms of physical well-being, people who understand themselves have fewer accidental injuries and diseases and recover sooner from illness than those who do not. They have what Abraham Maslow (1962) long ago characterized as an "appreciation of the body," which leads, by extension, to the personality. They are the healthy, self-actualizing individuals who not only know what to do but, more important, are moved to do it. They are not indifferent to the consequences of their behaviors, but rather are sensitive to the fragility and dearness of life. They share a profound awareness of the potentialities of a fully awakened human being. They perceive a clear relationship between their well-being and the healthy development of the well-being of the society in which they live.

OUR TECHNOLOGICAL SOCIETY

Americans today live and interact within a highly technological environment. For one to be truly educated, one must be technologically literate. But beyond having the skills and knowledge to function in this "new" world, technology has had a negative impact on the way we move.

Computers, telecommunications, and an array of new manufacturing technologies have drastically increased industrial production, thereby reducing the need for human workers. Much of the movement workers once performed, termed large muscle activity, now can be done with the touch of a finger. The second industrial revolution, as it has been called, has had and will continue to have a profound social impact on men and women long associated with physical labor. The installation of an increasing number of robots in the workplace and the social costs of additional automation are of considerable significance to the educator. As automation continues to advance, the lifestyles of the workforce are subject to change. Are they prepared for less physical work on the job and possibly more leisure time?

With worker well-being becoming a national concern, a growing number of corporations have invested millions of dollars for the development and implementation of wellness programs. They were prompted to make this investment due to (1) diminished worker productivity, (2) increased absenteeism, and (3) increased health insurance and health care costs due to worker illness. Wellness programs typically exist in one of three forms:

- Work-site health promotion
- Work-site fitness
- Web-based fitness

All three programs are designed to help employees develop a wellness lifestyle. This is accomplished through a combination of the following services:

- Development of a personal wellness profile
- Health screening that includes cholesterol, body composition, and fitness assessments
- Classes dealing with such topics as stress management, nutrition, and smoking cessation
- Incentive programs whereby employees receive financial bonuses or prizes for practicing healthy behaviors
- A newsletter with wellness information

Work-site fitness programs generally provide exercise facilities as well as the above listed services. Facilities are often available to employees before, during, and after the work day.

Although people should not worry that technology will create a sick and impersonal society while robbing people of their freedom and humanity, it is nevertheless true that this possibility exists, especially if an educational effort to minimize its negative effects is not made. One task of both schools and communities, impacting more than a hundred million school children, is to prepare people for the world of automation and an increasing production-line way of life. Questions arising from this existence include the following:

- If approximately 15 percent of the nation's workers are in the so-called learned professions, what type of education should be advocated for the other 85 percent?

- How rich can we make the lives of the masses if the boredom of routinized automation is to be made bearable?

- How can we give people something to struggle for, master, and conquer so they can achieve and maintain their self-respect and dignity?

OUR MULTICULTURAL SOCIETY

Dante said that the worst place in hell is reserved for those who in times of great moral crisis take a neutral stand (Alighieri, 1948). Martin Luther King, Jr. stated that injustice anywhere is a threat to justice everywhere (King, 1987). Both of these statements suggest a need for multicultural education.

Baldwin (1989) explains that the purpose of education is to create individuals who have the ability to look at the world for themselves, to make their own decisions, to say that this is black or white, to decide whether or not there is a God in heaven. Asking questions of the universe and then seeking to answer them is the way a person achieves identity. However, some societies prefer individuals who are not independent thinkers; those who will simply obey a ruler. Yet, a society composed of such individuals will surely perish.

The cultural, ethnic, and racial diversity that was long ago global and not-so-long ago national, has now reached our schools, becoming local. It is estimated that by the year 2010, more than 40 percent of public school students will be minorities. With this ever-expanding diversity among students, it is now essential that they come to know, accept, and respect the cultural heritages of all people. In fact, our very survival has become more dependent on communication with and a better understanding of all cultures.

Such diversity poses a tremendous challenge for education as it strives to develop effective, productive members of society. As Butt and Pahnos (1995) stated, providing excellence and equity in education is difficult when both teachers and students have different means of communication, patterns of participation, and views of the world. This situation is compounded by the social and cultural changes occurring in today's world. To meet this challenge we must help students not only to know, but also to care, and ultimately to act. Chepyator-Thomson (1994) explains that as educators, we need to learn about increasingly diverse groups of people and, in turn, develop culturally sensitive programs to more effectively teach students from radically different cultural or social backgrounds. Educators need to provide all students with knowledge that enables them to care about people and social issues and moves them to improve society.

The foundation of multicultural education stems from the word *culture* and includes one's means of communication, language, beliefs, values, and attitudes; in essence, one's behavior (Tiedt & Tiedt, 1990). It is grounded in ideals of social justice, equity, and a dedication to providing educational experiences through which all students have an opportunity to reach their full potential. Within the physical education curriculum, this ideal is considered relative to program content, or the movement experiences provided. Through these movement experiences and the social and cultural interaction they provide, students can develop the interpersonal skills that contribute to mutual understanding and acceptance (Chepyator-Thomson, 1994).

OUR OVERPOPULATED SOCIETY

During the 2000 census, 281.4 million people were counted in the United States. This was a 13.2 percent increase from 1990. The population growth of 32.7 million individuals between 1990 and 2000 represents the largest 10-year increase in American history. The population within metropolitan areas increased by 14 percent. Social workers, public health officials, and recreation specialists are among those who believe that the growth of a city threatens the welfare of its inhabitants because it stifles individual and group mobility and expression. The question, then, is whether this dramatic increase in U.S. population will negatively affect our way of life. Is human welfare measurably affected by the density impact of a growing populace?

Population, or rather population density, must be viewed in terms of optimum population size as it relates to the quality of life. These expressions are somewhat subjective until associated with, for example, natural resource depletion, human nutrition, and psychosomatic illnesses, all of which are part of the stress syndrome that can occur as the result of overcrowding. The question as to what population size a city can comfortably manage and the earth can adequately support has become more urgent as population size has increased.

Significant from an educational viewpoint will be the difficulty of crowding more people into smaller spaces, and the real or potential increases in transportation demands, crime, air pollution, noise pollution, water pollution, drug addiction, and an array of other problems impacting one's physical and mental well-being. Consequently, the needs of urbanites and suburbanites will have to be carefully reviewed. The educational implications are numerous, including the need for more schools, teachers, and other resources to better meet the needs and interests of children. The role of the health/physical education/recreation specialist is relevant in bringing youth and adults together to find ways of enjoying forced proximity to each other. This is necessary if violence is to be reduced in cities.

As population density increases, privacy is threatened and humans become stressed, often times against unseen forces. Many people suffer

Increasing population size can stifle individual and group mobility and expression.

from apprehension and anxiety, fear and anger, and either resign themselves or rebel. Individuals who are determined, who know where they are going, will assert themselves in a number of socially accepted ways. However, when the need to be assertive is blocked, the stage is set for aggressive, antisocial behavior. When no outlet is available, violence often follows. Therefore, it is not surprising that physical activity through games, sport, and dance can provide a wholesome outlet for these aggressive feelings. James F. Conant (1961), a former president of Harvard University, wrote that if he were to name one educational program that potentially could do the most to reduce just one city problem—that of school dropouts—he would select physical education. Indeed, if quality physical education, intramural, and interscholastic athletic programs are available, the attention and energy of young people can be channeled into these activities. This fact alone should lend support for such school programs.

As the need for physical activity and the space required for this activity increases, our cities will be unable to meet this demand. This shortfall will necessarily impact the adjoining suburbs. The problems of the city are no longer merely confined to just the city. It has become necessary to combine urban and suburban school boundaries, cutting across geographic, economic, and social boundaries.

OUR AGING SOCIETY

In 1950, 8 percent of Americans were 65 years of age or older. In 2000, 12.4 percent of the total population of the United States was in this age group. The number of Americans aged 65 and older rose from 28 million (one in ten) in 1990 to 35 million (one in 8) in 2000. This number is expected to double by the year 2050. The implications arising from this statistic are far reaching. Older people will be better educated, will have more vitality than those who preceded them, and will affect many aspects of society, from politics to health care.

A variety of human movement programs for this aging group currently exist throughout the country. In addition, physical activity has become a common prescription for people in long-term care facilities. Moreover, the AAHPERD Committee on Aging has been working for years with grant money from the National Institute on the Aging to update knowledge pertaining to exercise limits for older adults.

Although biological implications from aging occur, including reduced flexibility, thinning hair, and increased clouding of the eye's lens, sometimes society's time-honored myths and prejudices propel older people into early senility and late-life depression. The sad fact is that much of this could be prevented. When Americans clearly understand the role that physical activity can play in the aging process, they can begin to redefine aging. As a person becomes physically fit, mental fitness also is improved. As a result, an individual ends up with greater human dignity in all aspects of life. Studies by the Life Extension Institute indicate that three primary factors determine a person's happiness following retirement (see figure 1.1).

The leisure-time pursuits factor is particularly meaningful when a person considers the nature of a lifetime education in preparation for two or more decades of retirement. The happiest retirees are those who stay

Regular physical activity helps older adults maintain functional independence and thereby enhances their quality of life.

Figure 1.1
Factors Affecting Retirement Years

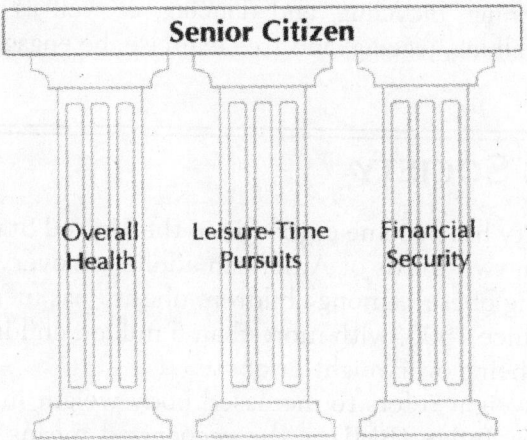

in their home communities and remain active through regular physical, mental, and social involvement. Moreover, weight control, daily physical activity, and adequate sleep appear to be highly beneficial. The secret of a rewarding life during retirement is for an individual to remain interested in what is taking place. Boredom alone can be reason enough for fatigue.

Older people continue to need physical activity, yet thousands do not begin to meet this need. Physical activity is important to help strengthen the heart and lungs, lower blood pressure, prevent obesity, and protect against the start of adult-onset diabetes. Regular physical activity can help older adults maintain functional independence and thereby enhance their quality of life. It is significant in preventing the common degenerative disease of osteoporosis, which affects women particularly after menopause. Bone stress brought about by physical activity is necessary even in older adults. Interestingly, large amounts are not required. It is the general stress applied to all parts of the bone framework that contributes to regeneration. It has been said that if physical activity could be packed into a pill, it would be the most widely prescribed and beneficial medicine in our nation.

Most people do not wear out; they rust out. In this respect, the playground movement champion, Joseph Lee, was accurate when he stated, "We don't cease to play because we grow old, we grow old because we cease to play" (Mero, 1909). Therefore, the way to stay active is to be active. Physical fitness and its components—muscular strength, muscular endurance, flexibility, and aerobic endurance—cannot be attained in any other way. However, only recently have older people in convalescent and retirement homes and in the expanding number of senior citizen centers been exposed to this fitness message and been provided access to fitness facilities.

The U.S. Department of Health and Human Services and the President's Council on Physical Fitness and Sports have promoted programs and implementation procedures for older citizens. These organizations have stressed that gardening, bicycling, and dancing, as well as physical activity in the form of walking, jogging, and free exercise, be engaged in on a daily basis.

OUR OBESE SOCIETY

Obesity has become epidemic in the United States. It is estimated that more than two-thirds of American adults are overweight, with one-half of these being obese. Among children, obesity has increased by more than 50 percent since 1990, with more than 5 million children between the ages of 6 and 17 being overweight or obese.

Overweight refers to increased body weight in relation to height. The Body Mass Index (BMI) is the recognized means of expressing this relationship. It is determined by dividing one's body weight in kilograms by the square of height in meters. An individual with a BMI of 25–29 is defined as being overweight and 30 or more is considered obese.

Being overweight or obese results from an energy imbalance that is the result of (1) eating too many calories and (2) not getting enough physical activity. The cause for this imbalance may be due to several factors. Three such factors include:

- Genetics
- Environmental factors
- Individual behaviors

Genetics and environmental factors may increase the risk of becoming overweight or obese, but individual behaviors, the choices one makes as to what and how much to eat and how much physical activity to engage in, have the greatest impact. Food options broaden daily with an ever-increasing array of prepackaged items, fast foods, and soft drinks. While such foods are convenient, they too often are high in fat and calories. As mentioned previously, technology has reduced the amount of physical activity expended in both our work and play hours, leading to an increasingly sedentary lifestyle. The President's Council on Physical Fitness and Sports (2004) states that those students most likely to become obese:

- Lack physical activity at home
- Lack physical activity at school
- Have a lower caloric expenditure
- Spend more time in sedentary pursuits (TV, Internet, video games, etc.)
- Eat larger portions during meals
- Consume a greater amount of refined carbohydrates

Why is this a concern? Excess weight negatively impacts the human body. Individuals who are overweight or obese have a higher risk for heart disease, hypertension, stroke, diabetes, and colon cancer. The medical cost for illnesses attributed to obesity exceeds $117 billion annually. The Surgeon General reports that poor diet and physical inactivity will undoubtedly overtake tobacco as the leading cause of preventable death in the United States (U.S. Department of Health & Human Services, 1996).

A PARTICIPATION PHILOSOPHY

What is needed today is a "participation philosophy." During the first half of the twentieth century, physical education philosophers such as H. Harrison Clarke, Charles H. McCloy, R. Tait McKenzie, Jay B. Nash, Frederick R. Rogers, and Jesse Feiring Williams stressed the need to educate youth to be participants in their world, not merely spectators viewing the activity and accomplishments of others. Their influence on the healthful practices of Americans was substantial. This educational philosophy is one path to a highly productive and fulfilled life. However, many Americans have moved away from an active lifestyle toward a more sedentary "watch and see" lifestyle.

Spectatoritis has long been a disease of Western civilization, so much so that "couch potato" has become a universally used term for one who is physically inactive. Too few people, especially young people, engage in physical activity. There are a variety of reasons for this, including (1) the changing family structure with children being raised by one parent or by two working parents; (2) the loss of the American "neighborhood," which provided a safe environment within which to move and play; and (3) a decreased emphasis on physical education in our schools. Without instruction in physical education children are less likely to develop the skills that enable them to be physically active. Ultimately, the quality of participation is highest when the participant has both the knowledge and skills pertaining to the various individual and group sports, dances, and other movement and fitness-related activities.

Following World War II, choreographer Rudolph Laban stressed the need to study and focus attention on the value of human movement in everyday life. Ten years later Dwight D. Eisenhower, moved by research findings indicating a lack of concern for the physical capacity of both youth and adults, established the President's Council on Youth Fitness, now known as the President's Council on Physical Fitness and Sports. A short time later John F. Kennedy (1960) presided over a somewhat sedentary population and made his famous remark, "We do not want a nation of spectators, but a nation of participants in the vigorous life."

Although the general public tends to agree that physical education should be an essential component of today's schools, physical education programs are nonexistent or inappropriate in some schools. Simply defining what is appropriate can be difficult. The American College of Sports Medicine (2000) considers appropriate physical activity to be that which involves large muscle groups in dynamic movement for periods of 30 minutes or longer, three or more days per week, performed at an intensity requiring 60 percent or greater of an individual's aerobic capacity.

In terms of increasing participation levels, there is much to be done. School physical education programs are designed to provide physical activity for all children. This participation in physical education might encourage extracurricular physical activity by children, leading to continued participation into adulthood. But only one-third of American children and adolescents in grades K–12 take part in daily school programs involving physical education. It is not any better outside of our schools. In today's society an inadequate level of physical activity is present among girls, women, older people, the physically and mentally disabled, and inner-city and rural residents.

HEALTHY PEOPLE 2010

In 1992, the U.S. Department of Health and Human Services (HHS) updated its previous report with the publication of *Healthy People 2000* (HHS, 1992), calling again for a public health revolution. In 2000, *Healthy People 2010* was published by the Office of Disease Prevention and Health Promotion (HHS, 2000). *Healthy People 2010* was developed through broad consultation with health experts, utilizing the best available scientific knowledge, and was designed to measure programs over time. This long-term effort identifies a wide range of public health priorities as well as two overarching goals, 10 leading health indicators (see table 1.1), and 28 focus areas (see table 1.2). The focus area where physical education has the most impact is number 22—physical activity and fitness. The specific objectives within this focus area are shown in table 1.3.

In 2004 the Centers for Disease Control and Prevention (CDC) released *Improving the Health of Adolescents and Young Adults: A Guide for States and Communities*. This document provides a framework for assisting state and local agencies in their efforts to establish priorities, develop intervention strategies, and measure progress toward improving the health, safety, and well-being of its young citizens.

Hippocrates taught that each disease of mind and spirit arises from a natural cause. Even one's values have much to do with both health-building and health-destroying behaviors. Too many Americans suffer from *value illness*, knowing what to do to become and stay well, but failing to do it. People know about the recovery power of rest but fail to get an adequate amount of sleep. They are well aware of the link between smoking and lung cancer, but they continue to smoke. They know alcohol affects one's ability to drive, but they continue to drink and drive. They know the role that regular physical activity plays in weight control and maintaining an appropriate level of fitness, but they do little to alter their sedentary lifestyle.

Because of the sedentary nature of our society, Americans have become health care dependent rather than health preventive; reactive rather than proactive. If we become ill, we go to our doctor to get well. This notion stems from a long-held belief that health services result in health improvements. But, we know today that this is not the case. Medi-

---

Table 1.1
Healthy People 2010: **Overarching Goals and Leading Health Indicators**

Overarching Goals

1. To increase the quality and years of healthy life for all
2. To eliminate health disparities among individuals regardless of race and ethnicity, gender, and income and education

Health Indicators

1. Physical activity
2. Overweight and obesity
3. Tobacco use
4. Substance abuse
5. Responsible sexual behavior
6. Mental health
7. Injury and violence
8. Environmental quality
9. Immunization
10. Access to health care

Source: U.S. Dept. of Health and Human Services. 2000. *Healthy People 2010.* Washington, DC: U.S. GPO.

Table 1.2
Healthy People 2010: **Focus Areas**

1. Access to quality health services
2. Arthritis, osteoporosis, and chronic back conditions
3. Cancer
4. Chronic kidney disease
5. Diabetes
6. Disability and secondary conditions
7. Educational and community-based programs
8. Environmental health
9. Family planning
10. Food safety
11. Health communication
12. Heart disease and stroke
13. HIV
14. Immunization and infectious disease
15. Injury and violence prevention
16. Maternal, infant, and child health
17. Medical product safety
18. Mental health and mental disorders
19. Nutrition and overweight
20. Occupational safety and health
21. Oral health
22. Physical activity and fitness
23. Public health infrastructure
24. Respiratory diseases
25. Sexually transmitted diseases
26. Substance abuse
27. Tobacco use
28. Vision and hearing

Source: U.S. Dept. of Health and Human Services. 2000. *Healthy People 2010.* Washington, DC: U.S. GPO.

Table 1.3
Objectives for *Healthy People 2010* Goal 22: Physical Activity and Fitness

22-1.   Reduce the proportion of adults who engage in no leisure-time physical activity.

22-2.   Increase the proportion of adults who engage regularly, preferably daily, in moderate physical activity for at least 30 minutes per day.

22-3.   Increase the proportion of adults who engage in vigorous physical activity that promotes the development and maintenance of cardio-respiratory fitness thee or more days per week for 20 or more minutes per occasion.

22-4.   Increase the proportion of adults who perform physical activities that enhance and maintain muscular strength and endurance.

22-5.   Increase the proportion of adults who perform physical activities that enhance and maintain flexibility.

22-6.   Increase the proportion of adolescents who engage in moderate physical activity for at least 30 minutes on five or more of the previous seven days.

22-7.   Increase the proportion of adolescents who engage in vigorous physical activity that promotes cardio-respiratory fitness three or more days per week for 20 or more minutes per occasion.

22-8.   Increase the proportion of the nation's public and private schools that require daily physical education for all students.

22-9.   Increase the proportion of adolescents who participate in daily school physical education.

22-10.   Increase the proportion of adolescents who spend at least 50 percent of school physical education class time being physically active.

22-11.   Increase the proportion of adolescents who view television two or fewer hours on a school day.

22-12.   (Developmental) Increase the proportion of the nation's public and private schools that provide access to their physical activity spaces and facilities for all persons outside of normal school hours (that is, before and after the school day, on weekends, and during summer and other vacations).

22-13.   Increase the proportion of work sites offering employer-sponsored physical activity and fitness programs.

22-14.   Increase the proportion of trips made by walking.

22-15.   Increase the proportion of trips made by bicycling.

Source: U.S. Dept. of Health and Human Services. 2000. *Healthy People 2010*. Washington, DC: U.S. GPO.

cal intervention is responsible for only a 3.5 percent decline in mortality since 1900. In addition, 80 percent of our health quality is affected by the environment, relationships, and the quality of education. For example, if cures were found for all forms of cancer, the average life expectancy would increase by only two years. On the other hand, a regimen of good nutrition, adequate physical activity, and appropriate health habits would increase average life expectancy by as many as seven years.

In a real sense, modern society is characterized by the spectacle of humans striving for perfection, yet knowing little about where they are headed. Efforts all too often fail to produce the peace of mind being sought. In an environment where speed, community status, and financial success are determinants of achievement, it is not uncommon to find individuals who cannot adjust to increasing pressures. Too many become over-fed and under-active. Others become overburdened with insecurity, anxiety, anger, and stress. Resulting tensions cause an array of maladies, including migraine headaches, indigestion, insomnia, irritability, and fatigue. It is becoming apparent today that this affects the school-age population as well as the adult population. A well-designed physical education program can do much to equip students with a defense against these pressures. The community-wide business of advancing physical education skills and knowledge is no small undertaking. It calls for a real understanding of the intricate relationship between unhealthy practices and disease as well as the health-related consequences associated with physical activity experiences. The following statistics emphasize the need for this relationship:

1. Cardiovascular disease affects more than 7 million Americans, causing nearly 1 million deaths each year; or one death every 32 seconds. Heart disease and stroke are the principal components of cardiovascular disease. They are the first and third leading causes of death for adults in the United States, respectively. They account for more than 40 percent of all deaths annually. Beyond the personal effects, the economic effects on our health care system are staggering. In 2003 the cost of heart disease and stroke was reported as more than $351 billion, which includes both health care costs and loss of worker productivity. Uncontrolled hypertension, obesity, and sedentary lifestyles only exacerbate this problem, which will continue to exact an enormous toll in lost lives and lost productivity, particularly if health and physical education efforts are taken lightly and subsequent improvements are left to medical practice alone (CDC, 2004b).

2. Following heart disease, cancer is the second leading cause of death in the United States. Death rates from some of the more common cancers, including lung, colon, breast, and prostate, have dropped in recent years. This is primarily due to early detection and advances in treatment. Still, the American Cancer Society (2004) indicates that cancer accounts for 1,500 deaths a day in the United States. More than a million new cancer cases will occur next year and one in three Americans will eventually have cancer. There is mounting evidence that everyday behaviors, including diet, tobacco use, and occupational stress, have much to do with this disease. Dietary modification, reduction of tobacco use, and an increase in physical activity to combat stress are, therefore, recommended strategies to reduce the incidence of cancer.

3. Injury or accidents are the leading causes of death among school-age children, claiming more deaths than all other causes combined. Within the general population, there were nearly 100,000 accidental deaths in 2002, with unintentional injury deaths up 2 percent from 2001. A fatal injury occurs every five minutes and a lifelong disabling injury occurs every 15 seconds (National Center for Injury Prevention & Control, 2002). Poisoning is second only to motor vehicles as the major cause of accidental deaths.

There are more than 15,000 poison and 44,000 motor vehicle deaths a year. There is a motor vehicle death every 12 minutes and a disabling injury every 14 seconds. There were more than 3,000 drownings and 150,000 swimming-related injuries in 2002. There is a considerable need for more safety instruction in all sports. In 2001 approximately 68 percent (170 million) of Americans participated at least once in some sport/physical activity as monitored by American Sports Data, Inc. (2004). Specifically, more than:

- 50 million individuals over the age of six were involved in such physical activities as running and cycling
- 40 million participated in recreational sports like basketball, softball, and tennis
- 15 million were outdoor enthusiasts involved in hiking, mountain biking, or skiing

Among this sport-related population more than 4 million injuries requiring an emergency room visit were reported, as well as an estimated 20 million less serious injuries.

4. In 2003, approximately 28 million American adults aged 18 and older received treatment for mental illness. The World Health Organization predicts that poor mental health will be second only to heart disease as the world's leading cause of death and disability by 2020 (Saraceno, 2002). It is estimated that from 10 to 12 percent of children and adolescents suffer from mental disorders. Suicide is the most serious outcome from these mental disorders. More than 5,000 suicides a year are committed by individuals under the age of 25. The contribution that stress makes to this unfortunate statistic is not completely understood. But according to some researchers, physical activity can produce short-term, if not long-term, relief (Taylor, Pietrobon, Pan, Huff, & Higgins, 2004).

5. The effects of drug abuse on our society are staggering, both on a personal and an economic basis. Results from a national survey conducted by the Substance Abuse and Mental Health Services Administration (2004) indicated that 19.5 million Americans aged 12 and older were currently illicit drug users. Marijuana was found to be the most commonly used drug, with 2.6 million new users in 2002. More than 50 percent of the population aged 12 and older, nearly 120 million Americans, drank alcohol. More than 32 million reported that they had driven at least once while under the influence of alcohol. Young adults, those age 18 to 25, were the most prevalent heavy drinkers. More than 70 million Americans use tobacco products. This is 29.8 percent of the population aged 12 and older. It is estimated that there are 2,000 new smokers under the age of 18 each day.

6. The fitness of both children and adults improved steadily from 1958 to the early 1980s. However, this trend is being reversed. Even though today's young children are among the most physically active segment of the population, the level of physical activity begins to decline as they

approach their teenage years and continues this downward spiral throughout adolescence (Brownson, Jones, Pratt, & Blanton, 2000). Among high school students:

- 35 percent are not physically active enough for aerobic benefits
- 45 percent do not play on a sport team during the school year
- 44 percent are not enrolled in a physical education class

It is clear that many individuals lack the physical capacity necessary to lead an active life. Our society must accept part of the blame for this, as we have become one that discourages physical activity. Our culture makes it *easy* to be sedentary and *not so easy* to be physically active. Reasons for this include the following.

- Our lives tend to revolve around the automobile. As a result, walking and bicycling have become a thing of the past.
- Heightened concern for personal safety has limited both the times and the places where children and adults can be physically active.
- Technology has provided myriad computer games that make sedentary activity enjoyable.
- States and independent school districts within states have reduced the amount of time students are required to take physical education.

The President's Council on Physical Fitness and Sports (1994) has demonstrated what can be done to improve fitness when adults, as well as children, are exposed to a quality physical education program that is carefully designed to meet individual weaknesses. Knowing there is a direct link between regular physical activity and improved health, more than four out of ten adults still indicate that they are not likely to increase physical activity in the near future.

In its purest sense, physical activity provides an opportunity for the individual to become refreshed, rejuvenated, and recreated. The concern in today's world is having the time to be physically active. Americans are free, without moral stigma, to use their time exactly as they choose. They have, indeed, been doing just that for years. Perhaps the ultimate question is whether the twenty-first-century American is wise enough to plan the use of nonwork time so that civilization will flourish. There is a need to educate Americans regarding the worthy use of leisure time. If this education is effective it may reap not only intellectual and physical satisfaction for the individual, but also creative, artistic, and spiritual satisfaction sufficient enough to contribute to an ultimate inner growth.

If physical education, or any other kind of education for that matter, is to contribute to the concept of equilibrium (a balanced existence) and reach the untapped resources of all human beings, it will have to bridge the gap between the working-stress world of the individual and the utopian state of meaningful recreational activity. In so doing, it will fortify the individual for both the world of work and the world of play.

Siedentop (2004) indicates that the wellness approach takes a holistic view, suggesting that a person's physical, mental, and psychological problems are all interrelated. For an individual to achieve wellness, that person must work, play, and socialize in a positive, balanced manner.

The need for quality elementary, middle, junior high, and high school programs of physical education exists side by side with the need for continuing education to address weaknesses in adult fitness practices. Major necessities include providing easily accessible paths for bicycling, jogging, and walking; offering greater opportunities for swimming; and making a stronger effort to disseminate information to the public on the value of physical activity for the total wellness of the individual. Americans need to become more physically active and steer away from the materialistic or comfort values of the *homo sedentarius* toward what has been called a *new feeling state*. Leonard (1975) expressed it well when he said, "We are discovering that every human being has a God-given right to move efficiently, gracefully, and joyfully."

INTERNATIONAL PHYSICAL EDUCATION AND SPORT

For several decades, a growing number of people have believed that physical education and sport can and should make a more effective contribution to the inculcation of fundamental human values everywhere in the world. In our vastly shrinking world, an increasing level of multicultural knowledge and understanding is essential. Through sport and physical education activities, countries can build better international friendships and understanding. This, in turn, could have a positive effect on international peace. Such a belief has the full support of the United Nations Educational, Scientific, and Cultural Organization (UNESCO). Moreover, the UNESCO Articles (1947) specifically call for programs designed to meet individual and social needs, with certain priorities given to disadvantaged groups in society. UNESCO has triggered worldwide efforts to expand opportunities in physical education. This is exemplified by its ratification of an International Charter of Physical Education and Sport. The charter indicates that the practice of physical education and sport is a fundamental right. More specifically, the freedom to develop physically, mentally, and morally through physical education and sport must be guaranteed within both the educational system and the various aspects of one's social life.

In addition to UNESCO, three other international organizations warrant discussion. AIESEP (Association Internationale des Écoles Supérieures d'Éducation Physique) or the International Association for Physical Education in Higher Education, established in 1962, is currently organized in 40 countries. The association's primary goal is to promote the integration of knowledge in physical education and sport and to provide scholarly opportunities for their application. More specifically its objectives are to (1) promote physical education and sport in higher education, (2) encourage the ongoing exchange of information among members, and (3) conduct research on new teaching methodology and evaluation techniques.

ICHPERSD (International Council for Health, Physical Education, Recreation, Sport, and Dance), first organized in 1958, is concerned with programs, policies, and educational aspects of its various allied fields. In addition, the council serves as a clearinghouse for information and ideas relative to these areas. It now boasts of a membership representing more than 145 countries.

ISHPES (International Society for the History of Physical Education and Sport) is the umbrella organization for sport historians throughout the world. It had its beginning in 1989. Its mission is to promote research and teaching in physical education and sport in order to facilitate and improve communication and cooperation between national and regional associations.

Web site addresses for these and other international organizations for physical education and sport are given in table 1.4.

Table 1.4
International PE and Sport Organizations

(AIESEP) International Association for Physical Education in Higher Education
http://www.zinman.macam98.ac.il/aiesep/aiesep.html

International Association of Physical Education and Sports for Girls and Women (IAPESGW)
http://www.udel.edu/HESC/bkelly/iapesgw/

International Council for Health, Physical Education, Recreation, Sport, and Dance (ICHPERSD)
http://www.ichpersd.org

International Council of Sport Science and Physical Education (ICSSPE)
http://www.icsspe.org

International Federation of Adapted Physical Activity (IFAPA)
http://www.ifapa.net

International Society for Comparative Physical Education and Sport (ISCPES)
http://www.iscpes.org

International Society for the History of Physical Education and Sport (ISHPES)
http://www.umist.ac.uk/sport/ishpes.html

THE ULTIMATE ENDEAVOR

The primary emphasis in this chapter has been to point out that the study of human functioning and well-being requires consideration of information from a broad spectrum of both medical and nonmedical fields. It is a multidisciplinary approach in which the physical education specialist has a substantial role to play. Ultimately, any proposed program of physical education should reflect this view. Moreover, such a program should make provisions for a coordinated effort in which the various medical and non-medical parties work together to maintain and advance human welfare and thus give the *celebration of life* concept unquestioned support.

If anything is apparent about the present condition of American adults, it is that the problems, diseases, and inadequacies that confront so many people did not suddenly appear. Rather, they emerged gradually, having been established during the elementary school years. Backaches, ulcers, hypertension, obesity, chronic fatigue, and the neurotic and psychotic behaviors related to such feelings as anxiety, worry, and jealousy are all tied directly to a pattern of living. An understanding of and an appreciation for the value of physical activity formed early in life, coupled with the proper skills and knowledge, set the stage for good health. The role that physical education can play in these formative years is immeasurable.

In the years ahead, the role of physical education as a vital part of the total education effort is of considerable consequence. A widespread understanding of the true nature and potentialities of physical education has never been fully realized. Some progress is being made as human movement is studied in its fullness and as the relationship between physical activity and wellness becomes more commonly acknowledged. With the determination and desire to work for quality programs, we may well usher in a golden era of physical education.

SUMMARY

1. Our hope is to live celebrated lives as awakened, enlightened individuals with the desire to make a contribution to the advancement of the world in which we live.

2. The true worth of physical education is best determined by how it affects the values of the individual members of the society in which it is taught.

3. Noted physical education philosophers, including Clarke, McCloy, McKenzie, Nash, Rogers, and Williams, stressed the importance of educating our youth to be active.

4. Technology has had a profound effect on the total well-being of the worker, creating a need for health promotion and work-site fitness programs.

5. Cultural diversity has resulted in a need for multicultural education. This diversity poses a challenge for all educators as they strive to develop students who care about people and are moved to improve society.

6. Physical education can play a role in lessening the problems brought on by both an increase in population and population density.

7. Physical education can play a vital role in increasing a person's longevity and happiness throughout the retirement years.

8. Physical education can play a vital role in combating the obesity epidemic in our society.

9. The U.S. Department of Health and Human Services report, *Healthy People 2010*, provides (1) data on our current health practices and (2) a strategy for improving our health.

10. Physical education has a lot to offer a society with a heightened need for leisure.

11. Physical education activities and events provide a medium to enhance international friendships and understanding.

QUESTIONS AND LEARNING ACTIVITIES

1. Does education have any responsibility for consciously changing our culture or do educators simply exist to perpetuate existing truths?

2. Interview one or more 80-year-old men or women. Ask the following:
   - How many hours a day did you work at making a living?
   - What new products or inventions impressed you the most during your teen years?
   - Who were your heroes as you were growing up?
   - What games did you play as a youth?
   - How did you use your leisure time as a youth?

3. In his utopian book, *The Shape of Things to Come*, H. G. Wells assumed that scientific thinking, modern engineering, and public education—by virtue of their intrinsic worth—would shape the kind of future of which an educated, middle-class citizen would approve. When he later wrote *Mind at the End of Its Tether*, he had become disillusioned as evidenced by his observation that Nazi Germany scored higher on scientific rationalism, engineering, and public education than did any other European nation. Take a moment to examine this observation. Are there implications here of any particular significance? Do educational goals and programs need careful definition as they relate to a civilization?

4. Provide a clear definition for value illness. What can physical education do to curtail the spread of this disease?

5. What can physical education do to help develop a culturally awakened individual?

6. In one sentence, provide your philosophy of physical education.

7. What can physical education do to combat the problems brought on by inner-city life?

8. Estimates are that 50 percent of American adults are obese. More than 25 percent of all children are obese. Unfortunately, obesity acquired during childhood or adolescence may well persist into adulthood, increasing the risk for some chronic diseases. Moreover, studies show that when these children are carefully observed, they are significantly less active than their nonobese peers. Many have little firsthand knowledge relating to what should be done about the obesity problem. What can physical education do to combat this epidemic?

REFERENCES

Alighieri, D. (1948). *The divine comedy, the inferno, purgatorio, and paradiso.* New York: Pantheon Books.

American Cancer Society, Inc. (2004). *Cancer facts and figures—2004.* New York: Author.

American College of Sports Medicine. (2000). *ACSM's guidelines for exercise testing and prescription.* Philadelphia: Lippincott, Williams & Wilkins.

American Sports Data, Inc. (2004). *A comprehensive study of sports injuries in the U.S.* http://www.americansportsdata.com

Baldwin, J. (1989). A talk to teachers. In R. Simonson & S. Walker (Eds.), *The Graywolf annual five: Multicultural literacy.* St. Paul, MN: Graywolf Press.

Brownson, R., Jones, D., Pratt, M., & Blanton, C. (2000). Measuring physical activity with the behavioral risk factor surveillance system. *Medicine and Sciences in Sport and Exercise, 32,* 1913–1918.

Butt, K., & Pahnos, M. (1995). Why we need a multicultural focus in our schools. *Journal of Physical Education, Recreation, and Dance, 66*(1), 48–53.

Centers for Disease Control and Prevention. (2004a). *Improving the health of adolescents and young adults: A guide for states and communities.* Atlanta, GA: Author.

Centers for Disease Control and Prevention. (2004b). *The burden of chronic diseases and their risk factors: National and state perspectives 2004.* Atlanta, GA: Author.

Chepyator-Thomson, J. (1994). Multicultural education: Culturally responsive teaching. *Journal of Physical Education, Recreation, and Dance, 65*(9), 31.

Conant, J. (1961). *Slums and suburbs.* New York: McGraw-Hill.

Dubos, R. (1981). *Celebrations of life.* New York: McGraw-Hill.

Kennedy, J. (1960). The soft American. *Sports Illustrated,* December 26, 15–17.

King, C. (Ed.). (1987). *The words of Martin Luther King, Jr.* New York: Newmarket Press.

Leonard, G. (1975). *The ultimate athlete: Revisioning sports, physical education, and the body.* New York: Viking Press.

Maslow, A. (1962). *Toward a psychology of being.* New York: Van Nostrand.

Mero, E. (1909). *American playgrounds: Their construction, equipment, maintenance, and utility.* Boston: The Dale Association.

National Center for Injury Prevention and Control. (2002). *Injury Fact Book 2001–2002.* http://www.cdc.gov/ncipc/

President's Council on Physical Fitness and Sports. (1994). American attitudes toward physical activity and fitness. *Journal of Health, Physical Education, Recreation, and Dance, 65*(1), 15.

President's Council on Physical Fitness and Sports. (2004). Seeing ourselves through the obesity epidemic. *President's Council on Physical Fitness and Sports Research Digest, 5*(3).

Saraceno, B. (2002). The WHO world health report on mental health. *Epidemiologia e Psichiatria Sociale, 11*(2), 83–87.

Siedentop, D. (2004). *Introduction to physical education, fitness, and sport*. New York: McGraw-Hill.

Substance Abuse and Mental Health Services Administration. (2004). *Overview of findings from the 2003 National Survey on Drug Use and Health*. Rockville, MD: U.S. Department of Health and Human Services.

Taylor, M., Pietrobon, R., Pan, D., Huff, M., & Higgins, D. (2004). *Healthy People 2010* physical activity guidelines and psychological symptoms: Evidence from a large nationwide database. *Journal of Physical Activity and Health, 1*(2), 114–130.

Tiedt, A., & Tiedt, I. (1990). *Multicultural teaching: A handbook of activities, information, and resources*. Boston: Allyn & Bacon.

United Nations Preparatory Educational, Scientific, and Cultural Commission. (1947). *Fundamental education, common ground for all peoples. A report to the United Nations Educational, Scientific, and Cultural Organization*. New York: Macmillan.

U.S. Department of Health and Human Services. (1992). *Healthy People 2000: National health promotion and disease prevention objectives*. Washington, DC: U.S. Government Printing Office.

U.S. Department of Health and Human Services. (1996). *Physical activity and health: A report of the surgeon general*. Atlanta, GA: Centers for Disease Control and Prevention.

U.S. Department of Health and Human Services. (2000). *Healthy People 2010: Physical activity and fitness*. Washington, DC: U.S. Government Printing Office.

2

EDUCATIONAL FOUNDATIONS

Outcomes

After reading and studying this chapter, you should be able to:

- Define
  Aerobic endurance
  Body composition
  A cultured person
  Education
  Flexibility
  Health-related fitness
  Intellectual competency
  Muscular endurance
  Recreational competency
  Social efficiency
  Strength
- Explain what it means to be a physically educated person.
- Identify the essential purposes/goals of physical education.
- Identify the five general objectives of physical education.
- Discuss both the physical and mental benefits of activity.
- Formulate a personal definition for physical education.

What special combination of knowledge, skills, and values should be developed in order to substantiate and give meaning to the celebrated life? To answer this question it is necessary to examine the educational fundamentals and determine to what extent they are made available to young people. More specifically, how is knowledge disseminated, how are minds liberated, how are skills developed, and how is the creative spirit encouraged to such an extent that the joy in movement has real meaning?

The physical education profession is currently making progress in a back-to-basics movement in middle and secondary schools by insisting that physical education is indeed an essential educational concern no less important than the fundamental processes of calculating and communicating. This is especially encouraging since Gallup and other poll-taking organizations have shown many times that a high percentage of parents know little about the public schools their children attend, and they know even less about the purposes of physical education in these schools. What it means to be physically educated is shown in table 2.1.

DEFINING THE ENDS

Educational objectives have meaning when there is a personal striving to make progress in a definite direction. To make progress, it is necessary to possess what Dewey (1970) referred to as a means of execution—that is, something has to happen for an objective to be realized. The word education embodies this idea since it comes from the Latin word *educare*, which means to "lead forth" or "draw out" the latent or potential qualities of a person. Education, therefore, is a process of changing behavior toward certain preconceived goals. The emphasis is on the process, which is not haphazard but rather is orderly and planned.

The essential purposes of education have changed somewhat in the last century. What Spencer (1860) wrote in the nineteenth century regarding the question "What knowledge is of most worth?" differs from modern ideas. According to Spencer, education was concerned with:

- Life and health
- Earning a living
- Family rearing
- Citizenship
- Leisure

A somewhat expanded list was put forth in *Cardinal Principles of Secondary Education* (Commission of the Reorganization of Secondary Education, 1918). These principles included:

Table 2.1
The Physically Educated Person

HAS learned skills necessary to perform a variety of physical activities

- Moves using concepts of body awareness, space awareness, effort, and relationships
- Demonstrates competence in a variety of manipulative, locomotor, and nonlocomotor skills
- Demonstrates competence in combinations of manipulative, locomotor, and nonlocomotor skills performed individually and with others
- Demonstrates competence in many different forms of physical activity
- Demonstrates proficiency in a few forms of physical activity
- Has learned how to learn new skills

IS physically fit

- Assesses, achieves, and maintains physical fitness
- Designs safe personal fitness programs in accordance with principles of training and conditioning

DOES participate regularly in physical activity

- Participates in health-enhancing physical activity at least three times a week
- Selects and regularly participates in lifetime physical activities

KNOWS the implications of and the benefits from involvement in physical activities

- Identifies the benefits, costs, and obligations associated with regular participation in physical activity
- Recognizes the risk and safety factors associated with regular participation in physical activity
- Applies concepts and principles to the development of motor skills
- Understands that wellness involves more than being physically fit
- Knows the rules, strategies, and appropriate behaviors for selected physical activities
- Recognizes that participation in physical activity can lead to multicultural and international understanding
- Understands that physical activity provides the opportunity for enjoyment, self-expression, and communication

VALUES physical activity and its contributions to a healthful lifestyle

- Appreciates the relationships that result from participation in physical activity
- Respects the role that regular physical activity plays in the pursuit of lifelong health and well-being
- Cherishes the feelings that result from regular participation in physical activity

Source: *Outcomes of quality physical education programs* by the Outcomes Committee of NASPE, 1900 Association Drive, Reston, VA 20191. Used with permission.

- Health
- Command of fundamental processes
- Worthy home membership
- Vocation
- Citizenship
- Worthy use of leisure time
- Ethical character

As far as health and physical education are concerned, both sets of purposes stress health as a primary aim and point to the need for education regarding leisure pursuits.

It is apparent that some of the first writers on education, including Socrates, Locke, and Rousseau, were just as enthusiastic about healthful living and its relationship to other educational objectives as were more recent educators, including James, Dewey, and Piaget. No matter how times change, individuals who are dynamically healthy are better able to satisfy personal needs and contribute to the welfare of society. The long-range goal of the physical educator, therefore, is to enable individuals to work toward the following objectives:

- Optimum organic health and the vitality to meet emergencies
- Mental well-being for dealing with the stresses of modern life
- Adaptability to and social awareness of the requirements of group living
- Attitudes and values leading to optimum health behavior
- Moral and ethical qualities contributing to life in a democratic society

LIFESTYLES AND EDUCATIONAL OBJECTIVES

In any era, the effort of realizing educational objectives must relate to the current and predicted lifestyles of the population. In the last several years some progress occurred in effecting healthful behavioral changes and in modifying lifestyles. As Dubos (1981) pointed out, humans can never adapt biologically to the diseases of civilization, but through creative adaptation their lives can be shaped through their responses to disease.

A person's success in adapting to environmental challenges, though remarkable, does have limits. Social, technological, and economic changes present novel challenges that sometimes exceed the adaptive capacities of most people. However, scrutiny of the aspects of one's daily lifestyle, individually and in combination with others, reveals that education can make a significant impact on a person's overall physical and mental capacity. Certain components of lifestyle, such as sedentary living and obesity, are now viewed as major risk factors. Moreover, mortality research indicates that nearly half of the ten leading causes of death in the United States can be traced to lifestyle. Fortunately, behavioral research has made it possible to understand behavior in the context of an individual's day-to-day activities. For example, getting a sedentary person to be physically active is more a matter of behavioral modification than of applying traditional medical practice.

Although physicians are being encouraged by both medical schools and associations to look more carefully at individual lifestyles, their limited and somewhat sporadic contact with people can have little long-lasting effect. Increasingly, the job of guiding individuals into more healthful living patterns will fall to health and physical educators in the schools and communities.

To modify or change behavior, the student must be viewed as a unified whole and not a conglomeration of parts. The whole person must be considered in order to effect a positive change in lifestyle and well-being. Decades ago both Jesse Feiring Williams and John Dewey championed this view. Oberteuffer (1965) indicated this was the crucial issue for physical education when he asked:

> What meaning does it have? What experience does it offer in the direction of that meaning? Does it make a contribution to man's entirety? . . . To survive, the physical education programs of the day must recognize the totality of man and be constantly mindful that man lives in a social setting, not in isolation.

Glasser (1976) refers to a poor-quality lifestyle as a condition that occurs when people lack the vitality and strength to find the happiness enjoyed by others. Such people lack the perception, energy, and will to overcome difficulties and pursue a lifestyle that includes fulfillment, pleasure, recognition, and a sense of personal value. Individuals need to strive for positive additions that will strengthen them and make their lives more satisfying. This is what Glasser has termed the *positive addition* theory. This theory explains why runners only have to think about putting on running shoes to feel the kinesthetic pleasure of movement.

PHYSICAL EDUCATION OBJECTIVES

Physical education concentrates on both the art and science of human movement. However, the ultimate objective is to employ movement as a means of contributing to the physical, mental, and social goals of education. Thus, physical education may properly be defined as education through physical means, primarily through large-muscle activity. It is learning to move by moving.

The five main objectives of physical education are the development and maintenance of:

• Health-related fitness

- Recreational competency
- Social efficiency
- Intellectual competency
- Culture

From time to time, teachers of physical education must be willing to scrutinize and carefully reappraise both the existing physical education objectives and the emphasis that is given to these objectives. Perhaps the matter of emphasis needs more attention. When reviewing the history of physical education, it is clear that some objectives were neglected in the past while others were overemphasized. Periods of great attention were given to (1) posture and mental health, (2) character development through sports participation, (3) body awareness through movement exploration, and (4) physical fitness.

At various times, professionals have asked if physical education should have a hierarchy of objectives; that is, should these objectives be prioritized? Such a hierarchy could bring additional attention to those objectives that appear to be of major importance in a given year. But this might also mean that some objectives may always be overemphasized at the expense of others.

Ancient Greece, with its cultural emphasis on sport, provides an interesting example of a society that placed great value on physical education. Not only was sport more prominent in Greek life than in any other culture before or since, but as Siedentop (1971) indicated, competitive athletics were the central focus of the culture and a primary criterion by which civilization was distinguished from barbarism. Moreover, a large part of Greek life was centered in the *palaestra* (gymnasium). Greek youth received sports skills instruction from dawn to midday. This experience formed the core of Greek education, an education to fulfill all human needs. The values derived from competition were intrinsic, not extrinsic and material. The goal was embodied in the concept of *aerate*; that is, excellence in performance and noble behavior. The Greek poet Homer described this concept as he wrote about the lives, personalities, and deeds of the ancient heroes who pursued a life of just and righteous conduct (White, 1989).

HEALTH-RELATED FITNESS

Defined briefly, health-related fitness is the capacity for activity. It is a very important aspect in contemporary life. Being healthy is a positive and dynamic quality closely related to two variables—diet and physical activity. It is often referred to as organic vigor or vitality, the physical element of behavior that permits a person to be active. It is demonstrated through physical performance.

Although related to health in general, fitness is more specific when carefully evaluated. For example, several people may be thoroughly checked by a physician and found to be free from disease and defects. Yet these people may vary in the degree to which they can perform physically. Some may tire from walking a short distance. Others may run the same distance without being winded. The greater the fitness, the greater the physical endurance and precision of movement. With greater fitness, people perform more efficiently and recuperate more easily from fatigue. Undergirding fitness is an organic soundness consisting of five components (see figure 2.1).

Strength is the ability of a muscle to develop tension resulting in the force necessary to move an object through space. Strength is typically exhibited in the human body as (1) postural strength, the ability to stand upright against the force of gravity; (2) dynamic strength, the ability to manipulate the body while performing locomotor skills, including walking, running, and jumping; (3) ballistic strength, the ability to move the limbs through various ranges of movement while performing basic game skills, including throwing, kicking, and striking; and (4) isometric strength, the ability to stabilize the body, parts of the body, or objects against the pull of external forces. *Muscular endurance* is the ability of a muscle to sustain the force necessary to move an object through space repeatedly. Strength is necessary to perform the push-up movement, while muscular endurance is necessary to perform the push-up movement continuously. *Flexibility* is the ability to move a limb or body segment through its full range of move-

---

Figure 2.1
Components of Health-Related Fitness

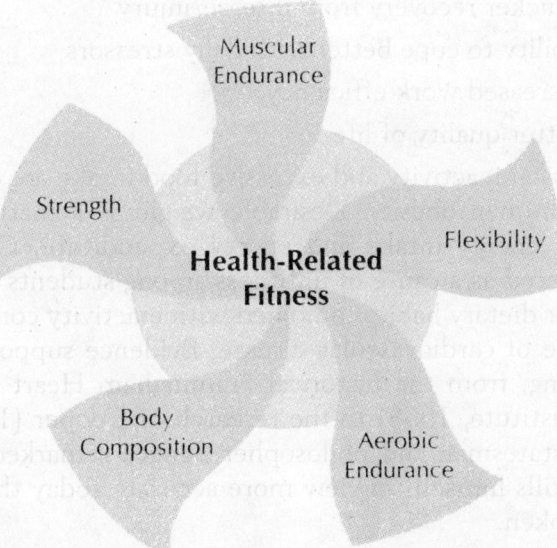

Muscular
Endurance

Strength

Flexibility

**Health-Related
Fitness**

Body
Composition

Aerobic
Endurance

ment. *Aerobic endurance* is the ability of the heart, lungs, and vascular system to provide sufficient amounts of oxygen and nutrients to the working muscle and carry the by-products of muscular work away from the working muscle. *Body composition* is the percentage of muscle, fat, bone, and other tissue in the body. The effectiveness of all body movements, whether large muscle or small muscle, depends on the status of these five, interrelated components.

In 1938 Conrad and Meister wrote that "Man is born; he struggles; he dies." This statement implies a pretty short life history. The key word in this remark is *struggles*. Life is a struggle. Therefore, the capacity to perform becomes key to the individual. With it, a person is able to strive and achieve a degree of happiness. This is accomplished not by acquiescing to what is, but by struggling for what should be; not by accepting but by questioning; not by receiving but by giving; and not by resting but by moving (Lawton & Rogers, 1937).

Physical activity is one way that our species makes itself known as a human animal. The ability to move is the foundation for physical education. Corbin, Lindsey, Welk, and Corbin (2002) emphasized the importance of physical activity when they presented its physical and mental benefits. Among these benefits are:

- A stronger heart
- A lower resting heart rate
- A lessened chance of heart attack
- A heightened chance of surviving a heart attack
- Enhanced sports performance
- Decreased susceptibility to disease
- Quicker recovery from disease/injury
- Ability to cope better with daily stressors
- Increased work efficiency
- Better quality of life

Physical inactivity and excessive food intake are etiological factors that lead to human obesity. Desirable weight is a matter of correct balance between energy intake and energy expenditure. Obesity is consistently encountered as a cause of unfitness among students and adults. Moreover, improper dietary habits combined with inactivity contribute to a rise in the incidence of cardiovascular disease. Evidence supporting this condition is convincing, from the historical Framingham Heart Study (U.S. National Heart Institute, 1968) to the research of Cooper (1978) and others. The Roman statesman and philosopher Seneca remarked that "Man does not die, he kills himself," a view more accurate today than when those words were spoken.

Working with the health teacher and others in school, the physical educator can do much to reduce the weight of an obese adolescent. Controlled physical activities like simple walking, jogging, and running can produce measurable reductions in percent body fat. Physical activity also aids the body's attempt to adapt to apprehension, worry, and anxiety by improving the sensitivity of the adrenal glands so they provide more steroids for counteracting stress. Physical activity improves functional efficiency at any age because the human biological design has not changed. Humans are meant to be active, not sedentary. As people grow older, fatigue is an obstacle to a happy life. Perhaps the best prescription for fatigue in older citizens, if not all citizens, is physical activity.

An abundance of evidence shows a significant difference between the longevity of individuals who are physically active and those who are not. Functional ability and physiological age can be extended for older people through controlled programs of such activities as walking, jogging, and swimming. As fitness improves, self-esteem rises and a stronger will to live emerges. This, in turn, reduces the mortality rate for a number of cardiovascular diseases and becomes an obvious contributor to longevity.

RECREATIONAL COMPETENCY

People are born for joy. In order to find life worth living people must have either joy in their lives or some hope of joy. It is joy that refreshes and recreates humans. The whole organism is involved. Recreation, therefore, is more than a game just for fun; it is an act of restoration. This restoration occurs through renewal of the physical, mental, and spiritual state and thus is more than a simple motor act. It involves deeper parts of the brain where the desire for divergence and refreshment is felt. Recreation can be viewed as an attitude. It is as akin to mental health as it is to physical well-being.

The recreation function is much more important than it may appear. In a world of increasing tension, both physical and psychological, an antidote must exist. Glands adjust to the constant demands of stress by releasing excess hormones for keeping the body in balance and striving for homeostasis. Without appropriate relief the body will suffer. Selye (1978) explains that the glands succeed for a while, but in the end the defense mechanism breaks down, arteries harden, blood pressure rises, and heart disease develops. Happiness is related to freedom from stress. An individual must have some respite from daily stresses, and recreation provides an excellent defense.

A variety of recreational skills, specifically those that are physical, need to be mastered early in an individual's life so participation and enjoyment can be extended throughout the lifetime. People tend to do those physical activities with which they have some expertise—some level of compe-

Recreation is an act of restoration that contributes to the individual's well-being.

tency. The paramount need, therefore, is to develop competency in a variety of skills. Poorly taught, poorly learned, and seldom practiced skills will rarely be incorporated into an adult lifestyle. A rewarding experience is one that leaves the participant in a state of mental and physical exhilaration, sufficient to promote peaceful relaxation and appreciation.

SOCIAL EFFICIENCY

Social efficiency is the ability to get along with others and exhibit desirable standards of conduct. It is a kind of social well-being related to mental and emotional health. School physical education can contribute measurably to the development of social efficiency by providing learning situations in which students can express themselves. This expression may occur through initiative, cooperation, leadership, followership, self-restraint, and loyalty to the group. A limited physical capacity sometimes handicaps an individual in pursuing social goals.

More than any other subject in the curriculum, physical education is organized to deal specifically with the elements of appropriate social behavior. Through carefully selected games, activities, and dances, students must assume a variety of responsibilities when working with others. Cooperation stems from a feeling for others or group consciousness. This awareness of others facilitates participation through both leadership and followership. Significantly, it is the freedom that individuals can exercise within the rules of the game situation that gives meaning to sportsmanship. This may be one of the most important concepts in education because without rules, freedom is meaningless; and without freedom, there is no long-term survival. In this respect, the sports movement has much to offer. It gives participants a degree of freedom along with the expectation that they will exercise proper restraint and adhere to the rules of the game. Moreover, the power of the word *sportsmanship* is universal. It is probably one of the clearest and most popular expression of morals today.

Sport and play have deep roots in culture and humanities simply because both provide life enrichment and fulfillment opportunities. Sport has had a profound impact on civilization for years. Sport has proven to be a means of allowing us to better understand not only how people live and work but also how they think. People learn how to live with others through the sport experience. Sports are an education in themselves. Sports relate directly to play and so they become a means of escape from the stresses of work and routine.

Widely considered a social institution of considerable magnitude, sport functions as something more than mere entertainment. It is now necessary to educate those who participate as well as those who spectate because the number of people who do both is increasing. Participation in organized sports through community park and recreation programs, YMCA/YWCA programs, schools, and sports clubs is extremely popular today. Education can help prepare for as well as promote a lifestyle that includes suitable sport experiences. Our society places a great deal of importance on sports. In fact, the term *sport sociology* has evolved because of this societal impact. A strong case has been made by several sociologists and historians for applying the traditions of the social sciences to the study of sport. Loy and Kenyon (1981) especially supported this idea. They define sport sociology as the study of sport in society as it affects human development, forms of expression, value systems, and the interrelationships of sport with other elements of the culture. Based on this definition, how can sports education be anything but a prime concern to all educators?

Aside from examining the health and social aspects, Loy and Kenyon view the sports experience in terms of the independent dimensions of pursuit of vertigo, aesthetic experience, catharsis, and/or ascetic experience. These underlying pursuits contribute to the impact of the sports experience on an individual's personal growth. Somewhat risky to the participant, the vertigo event is provided through the thrill of speed, acceleration, sudden change of direction, or exposure to dangerous situations. The aesthetic experience is captured through several senses. Catharsis is achieved through the release of tension that builds with frustration. The ascetic experience is seen as the punishment a person endures in both competitive situations and extended training periods as the gratification received from achievement is delayed.

A positive relationship exists between an individual's personality traits and social adjustment and his or her level of skill in sporting activities. Individuals with high general motor ability are inclined to be more sociable, dependable, tolerant, competitive, and popular among peers. Boys with low general motor ability tend to have feelings of insecurity, difficulty in social relationships, emotional instability, and a negative self-concept. However, it is the opportunity to display one's general motor ability—so highly correlated with game skills—in a group sport situation that fosters efficient social actions.

Individuals simply have to make adjustments in order to function in a social group. Variables that influence group performance include the size of the group, its leadership, the nature of the task, the influence of affiliation needs versus achievement needs on the part of the members, and the cohesiveness of the group. Basically, this is the way participants learn more about themselves as they struggle for higher levels of performance. The spirit of the ancient Greeks is reborn whenever players come to grips with themselves while struggling for perfection. Thus, the athlete does not retreat from a challenge but rather meets it head on. This is the way a peak experience in sport is attained—an experience in which the individual has an involuntary, transient experience of being totally integrated, functioning fully, and in complete control of the situation.

The social efficiency objective takes on greater magnitude when the inner purposes of play are considered. Cowell and France (1963) summarized the values of play education. When asked why play is important to children, they responded that play:

- Is a wholesome safety valve of prehuman origin for aggressions and other drives
- Allows the organism to test not only its ordinary powers, but its originalities, before responsibilities are too critical
- Bears some relation to the business of life, being in some measure the childhood form of work
- Provides wholesome compensations for frustration and failure experienced in other areas
- Provides opportunities for creativity
- Satisfies psychic hunger for activity, achievement, belonging and recognition, and similar needs
- Affords the normal mechanism for release of the imagination and a legitimate means for needed occasional escape from reality

- Provides opportunities for experiencing thrills and successes as well as proper dosages of risks and failures that contribute to character-building
- Develops an individual's resources for effective adjustment to solitude
- Develops a give-and-take; a subordination of the self and a loyalty to the team, which are of great social value
- Has moral significance, providing for improvement of values concerning fair play, cooperation, and other social virtues
- Encourages attention and therefore personality integration since interest is inherent in the activity itself without extraneous or interest-distorting motivations

INTELLECTUAL COMPETENCY

One of the primary responsibilities of the physical educator is to find ways to motivate students to continue participating in physical activities long after they leave school. One approach is to provide an adequate knowledge and understanding of the important benefits of physical activity. Inherent in this approach is having an appropriate answer to the following common questions by students:

- Why do we need to be physically active?
- Why must we warm-up before playing?
- Why must we do drills?

As physical educators, we must have answers to these and other questions. Individuals tend to defend things they thoroughly understand; moreover, people tend to act in support of their convictions. Today's students need to (1) understand why physical activity is necessary and why physical skills are beneficial and (2) appreciate the unique role of recreational activity in providing for a rich and full life. This is one way to meet the intellectual competency objective. A second method is to teach cognitive as well as the psychomotor content. An understanding of cognitive content, including game rules, terminology, and strategy, is essential if students are to fully appreciate the activities and sports in the curriculum. Generally, this can be accomplished using such techniques as lectures, videotapes, and independent study.

Vigor and intellectual growth are natural allies. If students understand not only the *what* and *how* but the all-important *why* of physical education, they—as future parents—will be more apt to lend enlightened support to our programs in years to come.

CULTURE

Perhaps the least understood objective of physical education is the development of the cultured person. Culture is not something ethereal but rather a concept that can be grasped. It is practical, involving a deep appreciation for life's activities, and is closely associated with a rich and full life.

Teachers of physical education who develop appreciation for rhythm and music through specific sport skills and dances and who employ form and color in creative activities are helping to develop the cultured person. Moreover, every teacher of physical education who successfully teaches a student a useful motor skill is contributing to the cultural objective. The meaning of culture substantiates this process because culture implies enlightenment, a refinement of thought. The mark of a cultured person is a refinement in both mental and moral powers as a result of particular training and its subsequent enlightenment. The cultured person feels a concern for many things. When the product of learning enriches and fills one with an appreciation of the arts, educators have succeeded in developing a cultured person.

Physical education can enhance the development of culture through the creation of the kinesthetic sense, the consciousness within the body for specific skilled movements. The appreciation developed through the kinesthetic sense is an artistic experience and is comparable to other artistic qualities. Accomplishing a difficult skill or gradually learning a new skill involves the same neural pathways and results in a neural pattern no less important to artistic appreciation than stimuli received through other senses. Listening to a symphony may cause a person to close his/her eyes and relax the body in the fullest appreciation. Art lovers may be completely lost in thought as they view a gallery masterpiece. Likewise, an athlete may hit a golf ball so perfectly or serve the tennis ball so accurately that a feeling of supreme satisfaction is experienced. The reactions of both art and music lovers are not unlike those of movement lovers. These reactions are outward manifestations of the cultured person. All three—art, music, and movement—have the capability of developing a person's fine artistic sense.

The inner feeling produced by these experiences is what counts in a discussion regarding culture. Music expresses the inexpressible. The same is true when describing the feeling one gets from viewing a highly skilled movement. The pure beauty of a thoroughbred turning at the post, the gymnast balancing motionless on the beam, a wide receiver coming down with the ball as both feet touch the ground just inside the white line, all bring culture to the heart of the spectator. Just as it is easier to describe the symphony performance than the actual music, it is likewise easier to describe the spectators at a football game or tennis match than the rousing impact of the physical activity itself. Both participants and spectators have experienced feelings that have cultural overtones.

SUMMARY

1. Education is a process of changing behavior toward preconceived goals. It is essential in an educational setting that this process be orderly and well planned.

2. Physical education has been defined as learning to move by moving. It is a medium that employs movement to enhance a student's total development; that is, physically, mentally, and socially.

3. The National Association for Sport and Physical Education has provided a thorough and explicit definition of a physically educated person.

4. Physical education instruction should be guided by five major objectives/goals that include the development and maintenance of health-related fitness, recreational competency, social efficiency, intellectual competency, and culture.

5. Fitness is a capacity for activity that has become an important aspect in today's society. One's ability to function through both work and play is dependent on the status of the five interrelated components of health-related fitness.

6. Recreational competency is achieved through the development of motor skills. The development of both the fundamental movement skills and the sport-specific skills allows the student to participate in and thereby enjoy a variety of physical activities.

7. The ability to get along with others and exhibit desirable standards of conduct leads to social efficiency. Physical education is an ideal medium for dealing with the elements of appropriate social behavior, among those being cooperation, self-restraint, and sportsmanship.

8. Providing knowledge about and an understanding of physical education is a way of developing intellectual competency.

9. Culture implies enlightenment, a refinement of thought. Physical education and sport activities provide an opportunity for students to develop an appreciation for movement. Furthermore, through structured activities, students can develop an appreciation for shape, form, expression, and creativity, which all contribute to the objective of developing the cultured person.

QUESTIONS AND LEARNING ACTIVITIES

1. As civilization becomes more complex, is there a tendency for education to become somewhat divorced from life? Is this true of physical education?

2. Formulate a list of contributions that physical education can make to a student's physiological, psychological, and sociological development. Be prepared to share this with others.

3. To better understand the concept of play, collect a number of definitions as expressed by philosophers, educators, therapists, and others. How do these definitions relate to play as an attitude? How do they relate to play as a means of education?

4. Human needs are not always accounted for in educational goals unless the planners of these goals are aware of both the aspirations and problems of humankind. Formulate a list of obstacles to achieving educational goals. Consider negative influences not only within the culture but also at the local school and community level.

5. Interview a number of physical educators in different schools and at different levels to determine what they believe to be physical education's general objectives.

6. Prioritize the chapter's five general objectives for physical education. Provide published support for your rankings.

7. What do you think are the three main purposes of today's schools? Attempt to provide published support for your answer.

8. What does it mean to be recreationally competent?

9. List specific ways that physical education can develop and maintain culture.

REFERENCES

Commission of the Reorganization of Secondary Education. (1918). *Cardinal principles of secondary education*. Washington, DC: Bureau of Education, Bulletin 35.

Conrad, H., & Meister, J. (1938). *Teaching procedures in health education*. Philadelphia: Saunders.

Cooper, K. (1978). *The aerobics way*. New York: Bantam Books.

Corbin, C., Lindsey, R., Welk, G., & Corbin, W. (2002). *Concepts of fitness: A comprehensive lifestyle approach*. New York: McGraw-Hill.

Cowell, C., & France, W. (1963). *Philosophy and principles of physical education*. Englewood Cliffs, NJ: Prentice Hall.

Dewey, J. (1970). *The way out of educational confusion*. Westport, CT: Greenwood Press.

Dubos, R. (1981). *Health and creative adaptation in the nation's health*. San Francisco: Boyd & Fraser.

Glasser, W. (1976). *Positive addiction*. New York: Harper & Row.

Lawton, S., & Rogers, F. (1937). *Educational paths to virtue, I*. Newton, MA: Pleides Company.

Loy, J., & Kenyon, G. (1981). *Sport, culture, and society: A reader on the sociology of sport*. Philadelphia: Lea & Febiger.

National Association for Sport and Physical Education. (1992). *Outcomes of quality physical education programs*. Reston, VA: AAHPERD.

Oberteuffer, D. (1965). *Background readings for physical education*. New York: Holt, Rinehart & Winston.

Selye, H. (1978). *The stress of life*. New York: McGraw-Hill.

Seneca, L. (1908). *Selected essays of Seneca and the satire on the deification of Claudius*. New York: Macmillan.

Siedentop, D. (1971). Differences between Greek and Hebrew views of man. *Canadian Journal of History of Sport and Physical Education, 2*, 30–49.

Spencer, H. (1860). *Education: Intellectual, moral and physical*. New York: Appleton Century Crofts.

U.S. National Heart Institute. (1968). *The Framingham study: An epidemiological investigation of cardiovascular disease*. Bethesda, MD: Author.

White, F. (1989). *The complete life of Homer*. London: Bell & Sons.

3

THE STUDENT AND CURRICULUM OBJECTIVES

Outcomes

After reading and studying this chapter, you should be able to:

- Define

  At risk

  Behavioral objective

  Curriculum

  General objective

  Hidden curriculum

  Outcomes-based physical education

  Specific objective

- Explain what is meant by the physical education potential.
- Understand and apply the taxonomies of educational objectives proposed by Bloom, Hastings, and Madaus.
- Distinguish the differences between general, specific, and behavioral objectives.
- Identify the basic components of a well-stated behavioral objective.
- Explain the basic school-level curriculum models.
- Describe each of the ten physical education curriculum models presented.
- Explain what the hidden curriculum is and how it can impact students.

I n hundreds of school systems around the country a substantial amount of change and innovation has occurred, the outgrowth of an admirable effort to make the whole process of schooling more pertinent to the needs of society and more vital to young people. Many of these changes have been accomplished precisely because of influential views from individuals like Thomas Hopkins (1941) and Jerome Bruner (1974). Their views stress the need for a sound philosophical base and clearly stated objectives for all educational programs. In physical education this practice has been demonstrated in numerous ways:

- Assigning more time to curriculum development
- Striving to meet student interests as well as needs
- Permitting individual student concentration in an activity
- Arranging personal conditioning programs
- Providing a broad variety of activities
- Instituting wider school-community programming of sports, dance, and recreational opportunities for people of all ages

Such an expression of concern for youth comes from a growing awareness of the intricate relationship between the feelings and aspirations of students and the program itself. In short, how do students view themselves? How do they view their school? How do they view their physical education curriculum?

THE STUDENT AND THE TIMES

The demand for accountability, a return to more traditional teaching subjects, and an increasing rise in enrollment coupled with decreasing financial support at the federal and the state levels challenge both school personnel and students to respond appropriately. Equally challenging is the lack of respect the teaching profession faces. What once was an admirable vocation is now viewed by some as what you do when you cannot do anything else. Whether the reason for this belief is due to attempts to increase taxes to support education or society equating an individual's importance to the size of his or her income, being a teacher is not an easy task. In addition, experience indicates that students have their own unique reactions to most educational pronouncements, programs, and processes based on their perceptions. To compound this, these perceptions may not always be well founded or rational. Yet these views are a major factor in determining how well students function in school.

Today's students are not the same as earlier generations of students because the world in which they live is not the same. Too many of today's children receive insufficient guidance. This is partly due to both parents working, one-parent families, parents whose lifestyle focuses less on their children, and the diminishing number of family-oriented neighborhoods that were once instrumental in raising everyone's children. Whatever the reason, the Children's Defense Fund (2004) presents some troubling data. In America, every:

- 6 hours a child dies from abuse or neglect (1,460 a year)
- 4 hours a teenager commits suicide (1,825 a year)
- 3 hours a child is killed by a firearm (2,920 a year)
- 19 minutes a baby dies before his or her first birthday (27,740 a year)
- 8 minutes a child is arrested for committing a violent crime (66,430 a year)
- 4 minutes a child is arrested for drug abuse (133,590 a year)
- 72 seconds a baby is born to a teen mother (432,890 a year)
- 66 seconds a student drops out of school (457,020 a year)

Students have always been influenced, to some degree, by the social and economic climate of the times. Students' lifestyles relate to current music, clothing, and entertainment trends. Moreover, students' lifestyles are impacted by the nature of their peer groups, the availability of recreational facilities, and the influence of church and other community groups. Students all too often engage in a conflict between parental control and an awareness of individual rights. As a result, many students develop attitudes and learn the types of skills that will prevent them from supporting social institutions as adults. The resulting behavior too often produces students who avoid responsibility, show disrespect for authority, and/or engage in what borders on criminal behavior. Some would argue that today's young people lack meaningful goals and that this deficiency most likely accounts for the unprecedented surge of social pathology in the United States today. The tragedy is that this situation is often preventable. These students have been called *at risk*. Although the meaning of this phrase varies, an at-risk student has unsatisfactory academic achievement and exhibits the potential to drop out of high school, become unemployed, and have a negative effect on society.

Peterson (1987) further defines this population by including students who are environmentally at risk. In other words, biologically and genetically normal students whose life experience and/or environment impose a threat to their developmental well-being, possibly resulting in low academic achievement, juvenile delinquency, drug abuse, and/or teenage pregnancy (Hellison & Templin, 1991). The proportion of students who are at risk is increasing. Responding to their needs must become a national priority. All educators, physical educators included, need to be aware of this at-risk population and work to develop and implement programs that can prove to be effective learning experiences.

Young people identify with programs that permit them to explore and feel the forces around them. From an early age they are ready to identify with people, situations, and the skills of cognition, sociability, and physical movement. Therefore, it is extremely important for the physical educator to meet this need for self-actualization by providing a curriculum that allows students to identify who they are. A child's self-concept (the value-free view of oneself), a child's self-esteem (the value one places on this view), and a child's self-confidence (one's belief in the ability to perform) all play a part in determining how that child will act and subsequently react. If children feel stupid they surely will present a noncomprehending face to the world. If children feel physically inept they will likely display a reluctance to move.

To fully understand young people, one must realize that they live in an intense present, a *here/now* environment. Much of what seems important lies either in the immediate life situation or in the near future. Youth rarely look into the distant future. To be effective, therefore, the physical educator must get close to the student's personal value system of *the moment*.

THE PHYSICAL EDUCATION POTENTIAL

Physical education (physical activity) was made for youth. It is the one subject in the curriculum that frequently appeals to large numbers of children, chiefly because of the chance to run, jump, skip, throw a ball, kick a ball, and express themselves through movement. It provides a means to challenge all students. Moreover, if physical education is properly presented it will bring people of many persuasions together. Because many kids actually enjoy physical education class, they may lose some of their hostility toward school in general.

Today's schools need organized, cooperative activities in which students can interact with peers. A curriculum properly conceived and content properly taught can promote group interactions and appropriate social behavior. In the school environment, the physical education program—perhaps more than any other program in the school—is where this process can readily occur. But for programming to be effective, it is important to remember that meeting students' needs is clearly an important factor.

FROM PHILOSOPHY TO OBJECTIVES

Fundamentally, it is the responsibility of every present-day teacher to think about both the ultimate purposes of and the existing practices in education. Philosophizing is an earmark of a professional.

Physical educators, ranging from the department supervisor to the curriculum coordinator to the in-class instructor, contribute to the overall philosophy of the program by evaluating diverse viewpoints. This evaluation is as germane to the development of the curriculum as is the availability of facilities and equipment. In fact, to rely solely on others for planning and preparing a list of desired student objectives, outcomes, and curriculum content for meeting those objectives is to miss experiencing one of the more challenging aspects of teaching. In addition, every teacher needs to be able to establish goals. Through this process of conceiving and implementing a program, teachers can achieve the personal satisfaction and fulfillment that draws many to the profession. Ideally, then, the supervisor and curriculum coordinators are only facilitators in the process of program change. What happens beyond that is up to the individual teacher.

A number of authors and philosophers in the physical education field have contributed to the discussion of objectives. They strongly support the rationale that objectives must ultimately relate to the potential meaning they have for the student. Nixon and Jewett (1980) express their objectives in terms of three clusters—fitness, performance, and transcendence. These refer, respectively, to physical condition, skill level, and psychological characteristics including self-awareness, heightened perception, kinesthetic discovery, self-mastery, creative expression, and joy of movement. After examining the fundamental movements and other content associated with physical education, Annarino, Cowell, and Hazelton (1986) divide objectives into five traditional categories of emotional responsiveness, interpretive and intellectual development, neuromuscular development, organic power, and personal-social attitudes and adjustment. These more traditional objectives are presented in an informational manner in table 3.1.

In their discussion of objectives, Cheffers and Evaul (1978) recognized such underlying concepts as how a student perceives, patterns, varies, adapts, improvises, and refines. They categorized the scope of human movement as that endeavor which pertains to how an individual adapts personal movement to *fit in* with the environment or *to change* the environment to suit personal needs and interests. Bloom, Hastings, and Madaus (1971) emphasized that all fields of education should subscribe to the cognitive objectives. Briefly, the six cognitive objectives can be applied to physical education as follows:

1. **Knowledge:** The student is able to recall specifics—methods, processes, theories, structures, or settings. (Example: Understands sport-specific terminology, history, and rules.)

2. **Comprehension:** The student is able to make use of something without necessarily relating it to other things; can demonstrate the ability to translate or paraphrase communication; the lowest level of undertaking. (Example: Explains the meaning of such items as cardiovascular endurance, aerobic activity, and sportsmanship.)

Table 3.1
Traditional Physical Education Objectives

1. **Emotional responsiveness**—an attempt to have students express joy at participation in games and sports; accept challenges that mean overcoming difficulties; derive enjoyment from a cooperative experience; and develop an increased appreciation of the aesthetic experiences inherent in all activity, including games, sports, and dance.

2. **Interpretive and intellectual development**—an attempt to encourage students to approach whatever they do with both imagination and originality.

3. **Neuromuscular development**—an attempt to develop skills, grace, and a sense of rhythm.

4. **Organic power**—an attempt to strengthen muscles, develop resistance to fatigue, and increase aerobic efficiency; the ability to maintain adaptive effort.

5. **Personal-social attitudes and adjustment**—an attempt to place students in situations that encourage self-confidence, sociability, initiative, self-direction, and a feeling of belonging.

Source: Annarino, A., Cowell, C., & Hazelton, H. (1986). *Curriculum theory and design in physical education.* Long Grove, IL: Waveland Press.

3. **Application:** The student is able to employ technical principles or abstractions, ideas, and theories in some way; can use information in a concrete situation. (Example: Observes a sporting situation or motor skill and clearly indicates how it may or may not be representative of what is expected.)

4. **Analysis:** The student is able to examine an idea, concept, or structure by breaking it down into its component parts so that the relationship between the parts is clear. (Example: Responds to a certain game situation by breaking down the intricate structure and patterns leading to successful participation.)

5. **Synthesis:** The student is able to bring together all parts and elements to form a whole; can work with pieces and make arrangements in a manner to create a structure or pattern not there before. (Example: Creates a dance or gymnastics routine from a number of individual skills.)

6. **Evaluation:** The student is able to make judgments pertaining to the worth of ideas, techniques, and materials. (Example: Demonstrates the ability to differentiate between an effective and ineffective volleyball serve, particularly as it contributes to the overall game objective.)

The literature in the physical education field contains listings of objectives that would be suitable for giving direction to teachers and students alike. Frequently, these lists serve as indications for the function of physical education and therefore have some merit for school use.

Specific Objectives in Physical Education

In discussing objectives, it is customary to proceed from (1) the program's major aims or goals—called general objectives to (2) its specific objectives to (3) the behavioral objectives for the students. In curriculum development the specific objectives further define the general objectives and often are related to a specific activity or content area. Table 3.2 provides the five general objectives discussed in chapter 2 with examples of two specific objectives for each.

Table 3.2
Developmental Objectives of Physical Education

The development and maintenance of:

• Health-related fitness
  The student will improve aerobic endurance.
  The student will improve abdominal muscular strength and endurance.

• Recreational competency
  The student will be able to drive a golf ball.
  The student will be able to kick a soccer ball.

• Social efficiency
  The student will display sportsmanship.
  The student will demonstrate self-direction.

• Intellectual competency
  The student will understand the basic rules of basketball.
  The student will describe the task sequence for a tennis forehand.

• Culture
  The student will develop an appreciation for square dance.
  The student will develop an understanding of the place and importance of basketball in our society.

Behavioral Objectives in Physical Education

The objectives closest to the student are called behavioral objectives. Other published terms include performance objectives, terminal competencies, and summative expectations. In recent years, due to teacher accountability, considerable attention has been given to the writing of these objectives because they focus attention on behaviors. Anything observable can be stated in a behavioral objective form and can be set forth in terms of action words. These observable acts include all behaviors in the three learning domains, whether psychomotor, cognitive, or affective. Examples of key words that can be used to describe a selected behavior are shown in table 3.3.

The distinct advantage of establishing student behavioral objectives is that from the beginning, the attention of the physical educator can be specifically directed toward helping students attain the prescribed outcomes.

A second advantage to writing clearly defined behavioral objectives is to allow the physical educator to be more selective in choosing course content, methodology, and resources necessary for meeting these outcomes. A

Table 3.3
Key Words Describing Behavior

analyzes	determines	interprets	runs
applies	develops	jumps	scores
appreciates	differentiates	kicks	swings
catches	discriminates	passes	throws
chooses	distinguishes	plays	values
dances	evaluates	relates	walks

well-written behavioral objective contains three basic components: (1) the observable behavior or task, (2) the criterion or performance level that indicates that the behavior has been completed, and (3) the condition(s) or situation under which this behavior is to be performed. The task component of a behavioral objective is written as a verb describing *what* the student is to do. Examples of tasks include dribble, catch, and define. The criterion component of a behavioral objective indicates the degree of acceptable behavior by describing *how well* the learner must perform the selected behavior. The criterion can be stated quantitatively, relative to such measures as time, distance, and percentages; or qualitatively, relative to form or technique. The condition component of a behavioral objective describes under what situation(s) the behavior is to be performed. It further defines when the behavior is to be completed. Examples of conditions are shown in table 3.4. Examples of complete physical education behavioral objectives in each of the three learning domains are contained in table 3.5.

OUTCOMES-BASED PHYSICAL EDUCATION

The determination of whether students have had a change in behavior as a result of exposure to a year of physical education is a measure of educational value. This measure helps satisfy the need felt by parents and the public at large for educational accountability. The desired change in student behavior has been approached recently from an outcomes-based approach. As Siedentop (2004) states:

> Teaching more carefully and more intensively to achieve outcomes is the big first step that physical education must take if it is to become better established in the school curriculum. (p. 285)

To achieve these goals two things must happen. Today's physical educator must:

- Carefully select the desired outcomes
- Teach to achieve these outcomes

Table 3.4
Condition Component of a Behavioral Objective

Behavior	Condition
Tennis forehand	When tossed from a pitching machine
Basketball free throw	In a game situation
Softball swing	From a batting tee
Soccer dribble	Around cones
Volleyball forearm pass	From a served ball

Table 3.5
Physical Education Behavioral Objectives

Psychomotor

- The student will dribble the soccer ball through the cone maze in less than 15 seconds on the 18th day of the unit.
- The student will score a minimum of 20 on the AAHPERD volleyball serving skill tests on the last day of the unit.
- The student will average 125 pins over the final three games bowled.
- The student will display correct form during the approach, hurdle, takeoff, flight, and entry phases of the forward dive.

Cognitive

- The student will score 75 percent on the tennis knowledge test on the final day of the unit.
- While viewing a videotape of an unskilled batter, the student will identify the three major errors of execution.
- The student will list all tennis court line markings when asked to do so.
- The student will describe the task sequence of performing an overhead volleyball serve.

Affective

- The student will exhibit sportsmanship in the basketball unit by indicating when a foul was committed.
- The student will choose a level of practice that is both appropriate and challenging for his/her ability level.
- The student will demonstrate self-direction by using pre-class time wisely.
- The student will assist a partner in developing proper technique, using Mosston's reciprocal teaching.

To guide the selection of outcomes, the National Association for Sport and Physical Education (NASPE) formed an outcomes committee in 1986 to (1) identify outcomes that define the physically educated student (see table 2.1) and (2) identify grade-specific competencies that serve as stepping stones, called *benchmarks*, for the physically educated student to achieve. NASPE (1992) published a list of suggested benchmarks for kindergarten and grades two, four, six, eight, ten, and twelve. Benchmarks for grades two, six, and ten are shown in tables 3.6, 3.7, and 3.8, respectively.

Table 3.6
Examples of Benchmarks—Second Grade

As a result of participating in a quality physical education program it is reasonable to expect that the student will be able to:

1. Travel in a backward direction and change direction quickly and safely without falling.
2. Travel, changing speeds and directions, in response to a variety of rhythms.
3. Combine various traveling patterns in time to music.
4. Jump and land using a combination of one- and two-foot takeoffs and landings.
5. Demonstrate the skills of chasing, fleeing, and dodging to avoid or catch others.
6. Roll smoothly in a forward direction without stopping or hesitating.
7. Balance, demonstrating momentary stillness, symmetrical and asymmetrical shapes on a variety of body parts.
8. Use the inside or instep of the foot to kick a slowly rolling ball into the air or along the ground.
9. Throw a ball hard demonstrating an overhand technique, a side orientation, and opposition.
10. Catch, using properly positioned hands, a gently thrown ball.
11. Continuously dribble a ball, using the hands or feet, without losing control.
12. Use at least three different body parts to strike a ball toward a target.
13. Strike a ball repeatedly with a paddle.
14. Repeatedly jump a self-turned rope.
15. Skip, hop, gallop, and slide using mature motor patterns.
16. Manage own body weight while hanging and climbing.
17. Demonstrate safety while participating in physical activity.
18. Recognize similar movement concepts in a variety of skills.
19. Identify appropriate behaviors for participating with others in physical activity.
20. Accept the feelings resulting from challenges, successes, and failures in physical activity.

Source: *Outcomes of quality physical education programs* by the Outcomes Committee of NASPE. Used with permission.

Table 3.7
Examples of Benchmarks—Sixth Grade

As a result of participating in a quality physical education program it is reasonable to expect that the student will be able to:

1. Throw a variety of objects, demonstrating both accuracy and distance (e.g., Frisbees, deck tennis rings, footballs).
2. Continuously strike a ball to a wall, or a partner, with a paddle using forehand and backhand strokes.
3. Consistently strike a ball using a golf club or a hockey stick so that it travels in an intended direction and height.
4. Design and perform gymnastics and dance sequences that combine traveling, rolling, balancing, and weight transfer into smooth, flowing sequences with intentional changes in direction, speed, and flow.

(continued)

 5. Hand dribble and foot dribble while preventing an opponent from stealing the ball.

 6. Consistently throw and catch a ball while guarded by opponents.

 7. Design and play small group games that involve cooperating with others to keep an object away from opponents (basic offensive and defensive strategy) (e.g., by throwing, kicking, and/ or dribbling a ball).

 8. Design and refine a routine combining various jump rope movements to music so that it can be repeated without error.

 9. Leap, roll, balance, transfer weight, bat, volley, hand and foot dribble, and strike a ball with a paddle using mature motor patterns.

10. Demonstrate proficiency in front, back, and side swimming strokes.

11. Participate in vigorous activity for a sustained period of time while maintaining a target heart rate.

12. Recover from vigorous physical activity in an appropriate length of time.

13. Monitor heart rate before, during, and after activity.

14. Correctly demonstrate activities designed to improve and maintain muscular strength and endurance, flexibility, and cardiorespiratory functioning.

15. Recognize that time and effort are prerequisites for skill improvement and fitness benefits.

16. Recognize the role of games, sports, and dance in getting to know and understand others of like and different cultures.

17. Identify principles of training and conditioning for physical activity.

18. Identify proper warm-up, conditioning, and cool-down techniques and the reasons for using them.

19. Accept and respect the decisions made by game officials, whether they are students, teachers, or officials outside of school.

20. Seek out, participate with, and show respect for persons of like and different skill levels.

Source: *Outcomes of quality physical education programs* by the Outcomes Committee of NASPE. Used with permission.

Table 3.8
Examples of Benchmarks—Tenth Grade

As a result of participating in a quality physical education program it is reasonable to expect that the student will be able to:

 1. Demonstrate basic competence in physical activities selected from each of the following categories: aquatics; self-defense; dance; individual, dual, and team activities and sports; and outdoor pursuits.

 2. Perform a variety of dance (folk, country, social, and creative) with fluency and in time to accompaniment.

 3. Assess personal fitness status in terms of cardiovascular endurance, muscular strength and endurance, flexibility, and body composition.

 4. Design and implement a personal fitness program that relates to total wellness.

 5. Participate in a variety of game, sport, and dance activities representing different cultural backgrounds.

 6. Participate cooperatively and ethically when in competitive physical activities.

 7. Identify participation factors that contribute to enjoyment and self-expression.

 8. Compare and contrast offensive and defensive patterns in sports.

(continued)

9. Discuss the historical roles of games, sports, and dance in the cultural life of a population.

10. Categorize, according to their benefits and participation requirements, activities that can be pursued in the local community.

11. Analyze and evaluate a personal fitness profile.

12. Use biomechanical concepts and principles to analyze and improve performance of self and others.

13. Appreciate and respect the natural environment while participating in physical activity.

14. Enjoy the satisfaction of meeting and cooperating with others during physical activity.

15. Desire the enjoyment, satisfaction, and benefits of regular physical activity.

Source: *Outcomes of quality physical education programs* by the Outcomes Committee of NASPE. Used with permission.

Following this publication, a Standards and Assessment Task Force was appointed to establish (1) content standards for school physical education programs that clearly identify what a student should know and be able to do as a result of being exposed to a quality physical education program, and (2) teacher-friendly guidelines for assessment of these content standards. The task force's work resulted in NASPE's 1995 publication, *National Standards for Physical Education: A Guide to Content and Assessment*. In addition, work has begun at the state level. Greenwood and Stillwell (1999) found that of the 42 state agencies providing curriculum guide materials, 39 provided outcomes for students. By establishing and monitoring these outcomes, physical education professionals gain credibility with the lay public, colleagues, administrators, and legislators.

THE NATURE OF CURRICULUM

Although the word *curriculum* has already been used several times in this chapter, little attempt has been made to examine its meaning. This is because a detailed discussion of curriculum is ideally preceded by a study of educational philosophy and objectives. Curriculum is derived from the Latin word *currere*, meaning to run. More common definitions of a curriculum include "a specified course of study," "all planned school activities," and "the body of courses offered by an educational institution." Both the derived word and these definitions seem to suggest an orderly plan and progression. An individual does not arrive at a schedule or course of study without engaging in some degree of organizing and planning.

In American public schools, curriculum is used as an all-inclusive term referring to the total program. All of the academic programs and the extra-class activities like band, student council, yearbook committee, intramurals, and interscholastic athletics are included in the curriculum and thereby are considered important. These activities, once called "extracurricular" (outside the curriculum), are now classified as cocurricular. The curriculum is a body of experiences that lies between the objectives and the teaching methodology employed for meeting those objectives (see figure 3.1).

A full program of activities will reflect the original aims and objectives. But its success with individuals will always depend on sound methodology, effective teaching, resources, and proper evaluation techniques. In short, the human factor—the teacher—has a lot to do with the achievement of curriculum objectives. It is possible to plan and develop a fine course of study only to find that it partially does the job for which it was intended because some teacher either failed to grasp its significance or was indifferent to its content.

An elementary school curriculum guide, for example, may give a detailed breakdown of student activities associated with the learning of specific game skills such as throwing, catching, and kicking. However, the teacher may teach the skills in a manner that does not involve the same experiences set forth in the curriculum guide. If, when the students are evaluated, it is clearly evident that they did not learn the skills, the teacher may be responsible because of the decision to ignore the guide. Obviously, some room for flexibility exists when following a prescribed course of study. However, physical educators who strive to teach within the framework of a well-designed course of study and periodically check student progress will generally contribute to the attainment of the program's objectives.

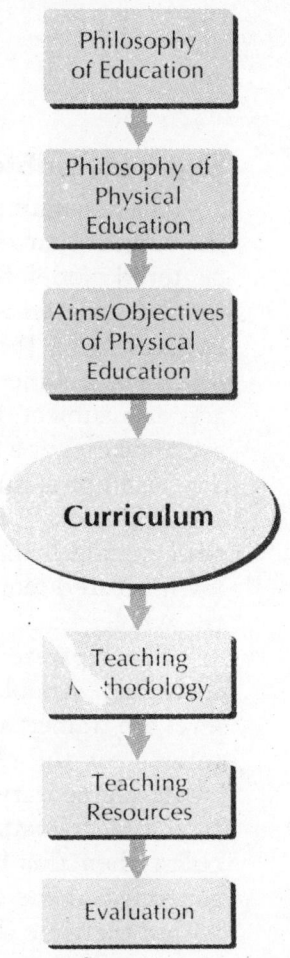

Figure 3.1
Central Position of Curriculum

Philosophy of Education

Philosophy of Physical Education

Aims/Objectives of Physical Education

Curriculum

Teaching Methodology

Teaching Resources

Evaluation

CURRICULUM MODELS

At the school level there are three basic curriculum models: separate subjects, humanistic, and broad fields (see figure 3.2).

Figure 3.2
School Curriculum Models

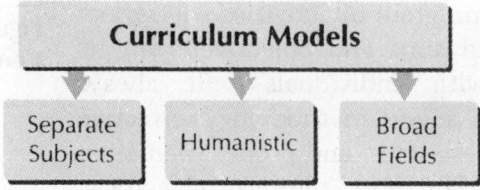

Separate Subjects Model

The separate subjects model is often called by other names, including the subject-matter model, the scientific subjects model, or the discipline-centered model. Students study each school subject (algebra, biology, etc.) for a designated amount of time each day. The emphasis is placed almost exclusively on the subject as a separate entity in the curriculum—not on science as a subject but on chemistry, physics, and biology; not on language arts as a subject, but on grammar, literature, and creative writing; not on social studies as a subject, but on history, geography, and sociology. Little, if any, attempt is made to relate one school subject to another. For example, physical education would be taught without any reference to health, wellness, or safe living. Within this model, there is a tendency for students to learn isolated facts and skills without seeing each content area as a part of the whole.

The separate subjects model does not generally create problems for secondary school physical education programs because teachers at that level are trained as specialists in a major field. It can, however, have implications for the elementary school physical education program. In many states elementary classroom teachers are prepared as generalists rather than subject-matter specialists. If physical education is not taught by a specialist then this responsibility falls on the classroom teacher. Too often, generalists have a limited physical education background. In addition, within the time allotment for a day, elementary teachers decide what and how much time to devote to each separate subject. Therefore, it is not uncommon for physical education to be either limited or ignored in both time and content.

Broad Fields Model

The broad fields model emerged from an awareness that students need balance between the many subject matters and the kind of understanding that allows them to make connections between varied content. In a limited sense, this is a curriculum approach in which subject-matter areas are grouped together under a common umbrella. For example, a course entitled "language arts" may include content dealing with reading, writing, literature, and oral communication; a course entitled "general science" may include content dealing with biology, chemistry, physics, and astronomy; and a course entitled "social science" may include content dealing with American history, political science, sociology, and geography.

In physical education, the broad fields model may include content dealing with personal health, wellness, and health-related fitness. Curriculum content could include:

- Assessment of the health-related components of fitness
- Development and maintenance of health-related fitness
- Knowledge of health problems associated with inadequate fitness
- Knowledge and application of biomechanical and physiological principles to improve and maintain fitness
- Knowledge of safety practices associated with fitness
- Knowledge of psychological values of fitness including stress management
- Nutrition information

Regardless of the subject area, this *total* approach more thoroughly allows the student to see, understand, and appreciate the relationships among several separate subjects learned under one general area. As the number of separate subjects is reduced, the number of teaching periods also is reduced, making it possible to lengthen the time of each period or to provide an opportunity for flexible scheduling. This allows the teacher, including the physical educator, to work without interruption to more fully develop the student's knowledge, skill, and appreciation of the subject matter.

Humanistic Model

Humanists share the view that education should contribute to the total well-being of the individual. Their philosophy is founded on the belief that humanism is a way of life centered on human interests and/or values. As a result, the humanistic model may emphasize an application for methodology more fully than for curriculum since it is grounded in the principle of respect for the individual. This model is supported because it focuses on what the students should *be* rather than on what they should *know*. The humanistic or existential curriculum reduces the importance placed on mastery of subject matter. Kelly and Melograno (2004) state that the emphasis is instead placed on (1) emotional concepts, including self-esteem, self-actualization, and self-understanding; and (2) social concepts, including cooperation, interpersonal relationships, and tolerance. Tomlinson (1999) calls this approach the *differentiated instruction* model. In this model the teacher begins at the student's level and builds on the student's intrinsic motivation to learn. Tomlinson highlights the importance of providing specific ways for each student to learn as quickly as possible, without assuming that one student's learning path is identical to another's. He believes that showing students that learning involves effort, some risk, and personal triumph can diminish the fear of failure, thereby encouraging participation.

Within the humanistic curriculum, students are not grouped for instruction chronologically but rather on ability. This nongraded system is designed to meet the student's needs by providing an opportunity for each child to advance at his or her own rate. Two implications arising from this approach are (1) determining how to ensure that the students are grouped properly and (2) planning to meet individual needs. If planning for the successful development of skills and knowledge is difficult, consider how much more difficult it is to plan for human interests and values. In any case, the humanistic model illustrates the extremely close relationship between the curriculum and the caring, personal methodology employed by the teacher. It is in this model that the true *art* of teaching is best realized.

The emergence of humanistic physical education is best characterized by a shift toward Maslow's concept of self-actualization. As such, movement exploration activities seem to be a means of achieving this end. In 1974, Siedentop stated that the humanistic model of physical education would eventually be applied by teachers who were selected more for therapeutic and social qualities than for knowledge. It was his contention that the development of sport skills would remain an objective, but would receive less of an emphasis. There is little evidence to support his prognostication. But, if physical educators chose to accept the humanistic model, their focus must shift to the affective aspects of physical activity.

In short, the affective domain is not to be overlooked. The humanistic emphasis has contributed to a shift from programs based on the competitive models for human striving and behavior to programs based on a more humane approach in which meaning, relevance, and ability are the primary values.

It is apparent today that the rights of the student determine, in part, the nature of the curriculum in all areas of education. Therefore, curriculum planning in physical education must consider the student's right to choose, right to safety, right to be informed, and right to be heard. In this respect humanism applies as much to sports and physical activity as to other scholastic subjects.

PHYSICAL EDUCATION CURRICULUM MODELS

At the departmental level there are a number of models or conceptual frameworks that can be used to meet the objectives discussed in chapter 2. Ten models are discussed below. These physical education curriculum models are designed to provide an effective means of selecting activities, developing teaching strategies, and providing movement experiences to meet the general objectives of the program.

Developmental Model

Siedentop (2004) believes that this curriculum model is, perhaps, the most applicable for physical education students in this century. Gallahue (2002) indicates that it is based on the concept that the development of a child's movement abilities occurs in distinct, sometimes overlapping, phases of motor development. This model is based on the belief that movement can be a medium used to attain developmental objectives. In using this model, the physical educator selects developmentally appropriate content (activities) to enhance both the rate and the quality of the student's development. Developmental physical education acknowledges the uniqueness of the individual and thereby is based on the belief that motor development is age-related and not age-dependent. The National Association for the Education of Young Children (NAEYC) defines a developmental curriculum model as one based upon knowledge of what is age-appropriate for students. The decisions on (1) what activities to include in the curriculum, (2) when to teach these activities, and (3) how these activities should be taught are based primarily on the activity's appropriateness for the individual student and secondarily on its appropriateness for a specific age group (Gallahue, 2002).

The implication for a physical educator planning to teach within the developmental model is to be familiar with the knowledge base regarding child development. Selecting and implementing a physical education program appropriate for a kindergarten class requires an understanding of the total (physical, motor, mental, and social) development of children at this age level. This will better allow for the selection of activities specifically geared to the learner's developmental level (Gallahue & Ozmum, 2002).

Movement-Education Model

The movement-education model, an alternative to the developmental model, had its beginning in the early 1970s. It uses Laban and Ullmann's (1960) analysis of movement classification, whereby movement themes are focused around such elements as body awareness, spatial awareness, and temporal awareness.

Movement education uses traditional content but employs a different process—how the content is taught—to achieve predetermined objectives. More specifically, movement education:

• Is student centered
• Involves exploration
• Is less formal than traditional physical education
• Involves problem solving and/or guided discovery
• Seeks to produce a feeling of satisfaction in movement

Movement education stresses body awareness, spatial awareness, and temporal awareness.

Logsdon and colleagues (1984) indicate that the purposes of movement education are to teach students to:

- Move skillfully, demonstrating versatile, effective, and efficient movement in situations requiring either planned or unplanned responses
- Become aware of the meaning, significance, feeling, and joy of movement both as a performer and an observer
- Gain and apply the knowledge that governs human movement

To this end the authors provide direction in using this model by stressing that physical education should be versatile and unplanned. Children need to learn to move with ease, but in doing so also need to learn how to cope with the unexpected in situations requiring movement, rather than being educated to rely on predetermined, prescribed motor responses.

Kirchner and Fishburne (1995) have found that movement education as an all-inclusive approach has proven to be most successful at the primary grade level. At the intermediate and secondary school levels, however, themes work best when integrated with other physical education curriculum models.

Health-Related Fitness Model

The five components of health-related fitness (muscular strength, muscular endurance, aerobic endurance, flexibility, and body composition) provide the content for this model's curriculum. The primary objective is twofold: to produce students who not only are fit but also know how to maintain that fitness. Smith (1993) states that physical educators utilizing this model should provide students with a lecture-laboratory approach. Specific content might include (1) lectures on health-related topics including drug abuse, cardiovascular disease, and the physiological basis of fitness and (2) laboratories, or practical lessons, using such fitness activities as jogging, Pilates, and weight training. Kelly and Melograno (2004) indicate that the curriculum should be designed around such topics as:

- The five health-related fitness components
- The five motor-related fitness components
- Diagnostic assessments
- The application of principles of training
- Nutrition, diet, and weight control
- Lifestyle management

Academic-Integration Model

The academic-integration model's primary focus is to foster the student's cognitive development. Content involves knowledge about the numerous subdisciplines of physical education including physiology, biomechanics, motor learning, sport sociology, sport psychology, and sport history. This model developed as the result of education reforms in the late 1980s, when attempts were made to make physical education more academic. AAHPERD demonstrated its support for this curricular approach with the initial publication of the Basic Stuff series in 1981 and again in 1987.

According to Siedentop (2004), this model has led to an emphasis on *integration*, where content traditionally taught in the classroom is now taught in the gymnasium. Its purpose is to combine traditionally diverse forms of knowledge (Placek, 1996).

Personal-Social Model

The personal-social model is an outgrowth of the humanistic education movement. The strongest advocate in physical education has been Hellison (1973, 1985, 2003). The ultimate goal of this model is to enhance the student's personal-social development by teaching both self and social responsibility. More precisely, the purposes are to help students:

- Cope within a highly complex world
- Achieve a higher degree of self-control
- Contribute more positively to the society in which they live (Siedentop, 2004)

This model was originally intended for at-risk students, but has applicability for students not considered to be at risk.

Hellison and Templin (1991) developed a two-part framework for applying this model: (1) empower students to assume more responsibility for their lives in a world presenting a variety of challenges and (2) teach students that they have a social responsibility to be sensitive to the rights, needs, and emotions of others. In so doing, students should:

- Feel empowered and purposeful
- Experience making responsible commitments to themselves and others
- Strive to develop themselves despite external forces
- Be willing to risk popularity in order to live by a set of principles
- Distinguish between their own personal preferences and activities that infringe on the rights and welfare of others

Specific goals within this framework are to:

- Develop sufficient self-control for respecting the rights of others
- Participate with effort
- Develop self-esteem
- Care about and help others

To achieve these goals, Hellison (1985) developed a six-level progression of socialization (see figure 3.3). Selected strategies are employed to help students interact as they work their way from one level to the next and make progress toward becoming caring individuals who ultimately possess and exhibit leadership qualities. These strategies include providing opportunities to:

- Learn the goals and the rationale for them
- Experience the goals

Figure 3.3
Hellison's Levels of Socialization

Leadership

Caring

Self-Responsibility

Involvement

Self-Control

Irresponsibility

- Make personal decisions
- Solve group problems
- Self-reflect

Sport-Education Model

The sport-education model has sport as the only content. According to Siedentop, Hastie, and van der Mars (2004), its intention is to help students

The goal of the sport-education model is to help students become skillful sports participants.

become skillful sports participants and sports persons. The teacher primarily acts as a facilitator by initially providing instruction in the specific sport and gradually empowering the students. To further this process, as many different aspects of sport as possible are incorporated into the physical education program. The true sport education model has five key characteristics:

- Sport seasonality—the yearly curriculum is organized by sport seasons rather than units.
- Team affiliation—students are members of a team and retain that membership throughout the sport season.
- Formal competition—competition is arranged in the beginning and utilized throughout the season (e.g., a round robin tournament).
- Culminating event—a winner is determined each sport season (e.g., a conference or state champion).
- Record keeping—scoring averages, assists, rebounds, etc., are kept and published to enhance interest and build tradition within this curriculum model.

Critics of this model contend that with the abuses evident in organized sports at virtually all levels, students should be exposed to fewer activities based on the professional sports model. However, proponents of the sport education model argue that since sport has gained such a position of prominence in our culture, it should be included in the physical education curriculum to enhance the development of appropriate sport behavior.

Adventure-Education Model

Adventure education has been incorporated into today's physical education curriculum due to the (1) increased interest in outdoor pursuits and (2) educational potential that exists within adventure activities. Siedentop, Mand, and Taggert (1986) explain that the purpose of these activities is for students to:

- Learn outdoor skills and enjoy the satisfaction of competence
- Live within the limits of personal ability related to an activity and the environment
- Find pleasure in accepting the challenge and risk of stressful physical activity
- Learn self-dependency with-in the natural world

Adventure education involves activities that take place within a natural environment or in a created environment, such as a ropes course. Students are challenged to complete a task and, in doing so, are required to overcome anxiety and/or function under stress. The focus is not on developing the skills necessary to complete the task or even on completing the task itself, but rather on the group involvement and specifically on cooperation, problem solving, and self-awareness. Adventure activities include backpacking, camping, canoeing, hiking, kayaking, orienteering, SCUBA diving, and wall climbing.

Adventure education activities can take place in either natural or created environments.

Multi-Activity Model

The multi-activity model is defined by the various activities offered within it (see table 3.9). The primary objective of the model is skill acquisition. To ensure that this objective is met, students are exposed to a variety of activities including team sports, individual-dual sports, adventure activities, fitness activities, and dance. Advocates think this diversity is necessary to meet the individual needs of students and to generate new physical activity interests. Critics argue that limited exposure to a wide variety of activities rather than extended exposure to a few activities will

Table 3.9
Multi-Activity Program Options

Team Sports	Individual Activities	Recreational Activities
Basketball	Aquatics	Angling
Football	Diving	Backpacking/hiking
Flag	Swimming	Bowling
Touch	Water safety	Camping
Hockey	Archery	Canoeing
Field	Badminton	Cycling
Floor	Conditioning	Dance
Lacrosse	Aerobic dance	Folk
Rugby	Aqua aerobics	Modern
Soccer	Circuit training	Social
Softball	Jogging	Square
Speedball	Slimnastics	Horseshoes
Team handball	Golf	Orienteering
Volleyball	Gymnastics	Rappelling
Water polo	Rhythmic	Sailing
	Traditional	Shuffleboard
	Handball	Skating
	Martial arts	Ice
	Aikido	Roller
	Judo	
	Karate	
	Paddleball	
	Pickleball	
	Racquetball	
	Self-defense	
	Skiing	
	Snow	
	Water	
	Squash	
	Table tennis	
	Tennis	
	Track and field	
	Weight training	
	Wrestling	
	Yoga	

result in a minimum level of skill competency at best. The selection of specific activities within this model is determined, in part, by student interest, the popularity of activities within the local community, and available resources.

Games-for-Understanding Model

The games-for-understanding model evolved from the early work of Almond (1983) and the studies of Bunker and Thorpe (1982). This model consists of various categories of games and sports that are selected from a games framework (Werner & Almond, 1990) and taught through a cognitive approach, utilizing guided discovery and divergent thinking (Mosston & Ashworth, 1994). This model allows the physical educator to provide students with the opportunity to solve strategic problems within game play. This is a shift away from the traditional approach of using more direct styles of teaching for developing sports skills.

To effectively apply this model, the physical educator must first select games from a classification. This classification system was proposed by Ellis (1983) and is outlined below:

Court	Field	Target	Territory
Shared	Fan shaped	Opposed	Goal
Divided	Oval shaped	Unopposed	Line

Court games involve a projectile being struck either in a shared court (handball and racquetball) or in a divided court (badminton and tennis). Field games involve opposing teams taking turns offensively and defensively on either a fan-shaped field (softball) or an oval-shaped field (cricket). Target games involve a projectile directed toward a target that is either opposed (shuffleboard) or unopposed (archery and golf). Territory games involve invasion by the opponent. They are divided into games in which either a goal is attacked (basketball, soccer, and hockey) or a line is attacked (football and rugby).

In the next stage, the game is played in a lead-up (conditioned) format. Students are guided to use correct strategies in each classified game by being asked specific questions. As the students become more strategic, the imposed game condition(s) are removed. Once the students are proficient game performers, they are encouraged to suggest game modifications and provide rationales for these modifications.

Lastly, students are grouped and encouraged to invent new games. As this is accomplished the teacher gradually moves from the role of facilitator, to advisor, to passive observer. Sensitivity to individual differences and student interactions is essential when grouping during the final stage. In addition, the physical educator needs to be cognizant of the discovery styles of teaching and prepared to teach in an environment where students are given the freedom to make decisions relative to learning.

Eclectic Model

The eclectic model, though probably not a true model, is included because it is the framework adopted by most schools. This model is comprised of two or more of the models previously discussed. It is widely used because schools are unlikely to adopt and adhere to one model for its total physical education curriculum or even the curriculum for one grade level.

Siedentop (2004) indicates that the eclectic framework works well in a program involving both required and elective content. For example, all students might be required to take a course developed within the health-related fitness model, but then would have the freedom to choose elective courses developed within the academic, sport education, and/or adventure education models.

The Hidden Curriculum

As stated earlier, it is essential to objectively present those behaviors that physical educators want students to exhibit. Doing so allows the physical educator to better select the content, methodology, and curriculum model for achieving these objectives. But a number of unplanned and unrecognized values exist that are taught and subsequently learned in a physical education environment. This has been called the *implicit curriculum* or *hidden curriculum* (Bain, 1975, 1976). Bain explains that students are repeatedly exposed to unconscious acts that are consistent in meaning. Examples of such acts include sex-role modeling by teachers, competition, and aestheticism (emphasis on physical beauty). When developing the curriculum and teaching the content within the curriculum, it is important for physical educators to be aware that this hidden curriculum exists and work to lessen its effect.

THE TEACHING DIMENSION

The value of any curriculum, regardless of the model(s) selected, depends on teacher effectiveness. It is beyond the scope of this text to delve into the procedures of how teachers should teach. This is the realm of a methodology text. It is unrealistic, however, not to comment on this topic since curriculum and methodology are so closely linked.

Carefully developed program content and quality instruction in physical education present a unified approach in which physical activity enhances motor development and advances social skills and self-concept. But the success of this approach depends on the variety and duration of many in-class behaviors. Psychologist Carl Rogers (1969) clearly stated that students will learn only what they want to learn and that teachers are first and foremost provocateurs who set the stage for student learning. Siedentop et al. (1986) assure that teachers are the backbone of education, the effectiveness of which lies in their day-to-day teachings. The level of effectiveness is judged by student performance. This basic assumption exemplifies the need for not only a meaningful curriculum with clearly defined objectives but also carefully selected teaching strategies.

SUMMARY

1. Today's curriculum reformers face the challenge of providing a program that will meet both the needs and interests of all students. This challenge is compounded since today's students are vastly different from those of a generation ago.

2. Physical educators need to be aware of the at-risk students and implement effective programs for meeting their special needs since physical education has this potential.

3. Objectives move from general to specific to behavioral. The objectives closest to the student are behavioral and, when stated, should contain the task, the condition, and the criterion.

4. The National Association for Sport and Physical Education has published a list of benchmarks (competencies) by grade level to clarify what it means to be physically educated.

5. The physical education curriculum includes the experiences that fall between the objectives and the teaching methodology.

6. At the school level, the curriculum can be organized using the separate subjects model, the broad fields model, or the humanistic model.

7. The developmental model of physical education is based on the premise that movement is a means of attaining developmental goals. The movement experiences included in this curriculum are selected relative to their ability to enhance the rate and quality of the student's total development.

8. The movement-education model has become the basic alternative to the developmental model, utilizing a different teaching process.

9. The health-related fitness model has gained renewed interest and is designed to produce students who are fit and know how to stay fit.

10. The academic-integration model, an answer to education reform, is designed to enhance cognitive development.

11. The personal-social model, which is humanistic in foundation, has the sole objective of enhancing the student's social development.

12. The sport-education model is designed to develop skillful sports participants.

13. The adventure-education model provides challenges to the student in either a natural or teacher-designed environment.

14. The multi-activity model has skill acquisition as its primary objective since students are exposed to diverse activities.

15. The games-for-understanding model is based on the premise that games should be taught through a cognitive approach.

16. The eclectic model is composed of two or more of the physical education models.

17. The hidden curriculum is composed of unplanned and/or unrecognized values that are taught in the physical education curriculum.

18. Just as the teacher is the key to curriculum development, effective teaching is the key to learning.

QUESTIONS AND LEARNING ACTIVITIES

1. Many professionals think a real need exists for improved accountability procedures in physical education. Some believe this can be accomplished by spelling out, in precise terms, exactly what behaviors are expected of the student. Do you agree? Are there any problems inherent in this approach? How else might we measure a teacher's accountability?

2. Write three specific objectives for each of the five general objectives for physical education.

3. Select a physical education activity and write a complete behavioral objective within the psychomotor, cognitive, and affective domains at the fifth-grade level. Select a different physical education activity and write a complete behavioral objective within the same three domains at the ninth-grade level.

4. Curriculum has been defined as being central to a student's education. Explain what is meant by this.

5. Interview a number of high school students. Find out what they like and dislike about their physical education programs.

6. Secure several recent state and/or local curriculum guides to see how their objectives are presented. Does each include general, specific, and behavioral objectives? Are concepts/outcomes/benchmarks more popular than objectives as a means of objectifying the program? Are there statements in the guide that attempt to alert teachers to how objectives are linked to evaluation practices?

7. If you were a third-grade physical educator, which physical education curriculum model(s) would you use? Why?

8. If you were a ninth-grade physical educator, which physical education curriculum model(s) would you use? Why?

REFERENCES

Almond, L. (1983). Games making. *Bulletin of Physical Education, 19*(1), 25–32.

American Alliance for Health, Physical Education, and Recreation. (1981). *Basic stuff series I*. Washington, DC: Author.

American Alliance for Health, Physical Education, and Recreation. (1981). *Basic stuff series H*. Washington, DC: Author.

American Alliance for Health, Physical Education, and Recreation. (1987). *Basic stuff series L*. Washington, DC: Author.

American Alliance for Health, Physical Education, and Recreation. (1987). *Basic stuff series II*. Washington, DC: Author.

Annarino, A., Cowell, C., & Hazelton, H. (1986). *Curriculum theory and design in physical education*. Long Grove, IL: Waveland Press.

Bain, L. (1975). The hidden curriculum in physical education. *Quest, 24*, 92–101.

Bain, L. (1976). Description of the hidden curriculum in secondary physical education. *Research Quarterly, 47*, 154–160.

Bloom, B., Hastings, J., & Madaus, G. (1971). *Handbook on formative and summative evaluation of student learning*. New York: McGraw-Hill.

Bruner, J. (1974). *The process of education*. New York: Vintage Books.

Bunker, D., & Thorpe, R. (1982). A model for the teaching of games in secondary schools. *Bulletin of Physical Education, 18*, 5–8.

Cheffers, J., & Evaul, T. (1978). *Introduction to physical education: Concepts of human movement*. Englewood Cliffs, NJ: Prentice-Hall.

Children's Defense Fund. (2004). *Key facts about American children*. Washington, DC: Author.

Ellis, M. (1983). *Similarities and differences in games: A system for classification*. Paper presented at the International AIESEP Congress.

Gallahue, D. (2002). *Developmental physical education for all children*. Champaign, IL: Human Kinetics.

Gallahue, D., & Ozmum, J. (2002). *Understanding motor development: Infants, children, adolescents, adults*. Boston: McGraw-Hill.

Greenwood, M., & Stillwell, J. (1999). State agency curriculum material for physical education. *The Physical Educator, 56*(3), 155–158.

Hellison, D. (1973). *Humanistic physical education*. Englewood Cliffs, NJ: Prentice Hall.

Hellison, D. (1985). *Goals and strategies for teaching physical education*. Champaign, IL: Human Kinetics.

Hellison, D. (2003). *Teaching responsibility through physical activity*. Champaign, IL: Human Kinetics.

Hellison, D., & Templin, T. (1991). *A reflective approach to teaching physical education*. Champaign, IL: Human Kinetics.

Hopkins, C. (1941). *Interaction: The democratic process*. Lexington, MA: Heath.

Kelly, L., & Melograno, V. (2004). *Developing the physical education curriculum*. Champaign, IL: Human Kinetics.

Kirchner, G., & Fishburne, G. (1995). *Physical education for elementary school children*. Dubuque, IA: Brown & Benchmark.

Laban, R., & Ullmann, L. (1960). *The mastery of movement*. London: McDonald & Evans.

Logsdon, B., Barrett, K., Ammons, M., Broer, M., Helverson, L., McKee, R., & Robertson, M. (1984). *Physical education for children: A focus on the teaching process*. Philadelphia: Lea & Febiger.

Mosston, M., & Ashworth, S. (1994). *Teaching physical education*. New York: Macmillan.

National Association for Sport and Physical Education. (1992). *Outcomes of quality physical education programs*. Reston, VA: AAHPERD.

National Association for Sport and Physical Education. (1995). *National standards for physical education: A guide to content and self-assessment*. St. Louis, MO: Mosby.

Nixon, J., & Jewett, A. (1980). *An introduction to physical education*. Philadelphia: Saunders.

Peterson, N. (1987). *Early intervention for handicapped and at-risk children*. Denver, CO: Love.

Placek, J. (1996). Integration as a curriculum model in physical education: Possibilities and problems. In S. Silverman & C. Ennis (Eds.), *Student learning in physical education: Applying research to enhance learning* (pp. 287–312). Champaign, IL: Human Kinetics.

Rogers, C. (1969). *Freedom to learn: A view of what education might become*. Columbus, OH: Merrill.

Siedentop, D. (1974). *The humanistic education movement: Some questions and issues in physical education and sports*. Mountain View, CA: National Press Books.

Siedentop, D. (2004). *Introduction to physical education, fitness, and sport*. New York: McGraw-Hill.

Siedentop, D., Hastie, P., & van der Mars, H. (2004). *Complete guide to sport education*. Springfield, IL: Human Kinetics.

Siedentop, D., Mand, C., & Taggart, A. (1986). *Physical education: Teaching and curriculum strategies for grades 5–12*. Mountain View, CA: Mayfield.

Smith, M. (1993). Utilizing different curriculum models to achieve the objectives of physical education. *Bulletin of Physical Education, 29*(1), 15–22.

Tomlinson, C. (1999). Mapping a route toward differentiated instruction. *Educational Leadership, 57*(1), 12–16.

Werner, P., & Almond, L. (1990). Models of games education. *Journal of Physical Education, Recreation, and Dance, 61*(7), 23–27.

4

CURRICULUM PLANNING IN PHYSICAL EDUCATION

Outcomes

After reading and studying this chapter, you should be able to:

- Define

  Public Law 94.142

  Student diversity

  Title IX

- Justify the need for a quality physical education curriculum.
- Identify the three categories of factors affecting curriculum development.
- Discuss the impact that Title IX has had on the physical education curriculum.
- Discuss the impact that Public Law 94.142 has had on the physical education curriculum.
- Identify the four central questions to answer at the beginning of the curriculum development process.

Ted Williams said:

I wanted to be the greatest hitter who ever lived. A man has to have goals—for a day, for a lifetime—and that was mine, to have people say, "There goes Ted Williams, the greatest hitter who ever lived." (Williams & Underwood, 1969, p. 7)

Ted Williams, at work, was an art form: his stance with the bat in his hand, the concentration, the swing, the follow-through—all were carefully engineered for the primary purpose of hitting a baseball out of the ballpark. For 19 major league seasons, Williams planned ahead by studying pitchers, arranging special practice sessions, and casting an eye for anything that might give him an edge. He was a superb example of how planning can help achieve a goal. It is through planning that a person's objectives are best attained. It is through planning that a physical education teacher selects appropriate activities and arranges them into harmonious, goal-reaching movement experiences. The physical education content is very important for developing students fully. "For harmony," said the Greeks, "is the music of the Gods."

Although the term is used to identify a field of study, curriculum is more appropriately used to illustrate a plan for the education of learners. Such a plan is founded on theory, research, and past professional practice. It is designed to achieve predetermined objectives. Without these objectives, planning may be limited and short-sighted. To prevent this, educators must engage in an orderly process of gathering, sorting, selecting, balancing, and synthesizing the relevant information from numerous sources.

Even the finest curriculum is subject to change. Publius Syrus said in 42 BC that "it is a bad plan that admits of no modification." Most of us know from experience that careful planning improves our likelihood of success in almost any endeavor; things left to chance seldom succeed. The Chinese philosopher Lao Tzu noted that in the affairs of men, there is a system. A reasonable plan for each community should be the immediate objective. Begin gradually, run pilot programs, solicit comments, evaluate results, and then with this background of experience the physical education effort can safely be extended.

CURRICULUM DEVELOPMENT

Curriculum has been a consideration of educators for centuries, but the specialized and systematic study of curriculum did not truly occur until the twentieth century (Kliebaard, 1968). The roots of curriculum development date back to the days of Johan Friedrich Herbert (1776–1841), a German educator whose ideas were widely accepted in America. Herbert taught that learning required an orderly attention to the selection and organization of subject matter. Moreover, he applied his views to what is now called physical education, as well as to other educational fields. In fact, he was one of the early philosophers to recognize the essential nature of a properly conceived and structured program of physical activity.

A curriculum in today's public schools is a product of socioeconomic forces. As a result it will find its fundamental philosophical purposes in the social and cultural sector in which it exists. The curriculum considerations of Hass (1987) included a concern for social forces as reflected in social goals, cultural uniformity and diversity, social pressures, social change, and planning. Since it is apparent that, as a society, people can be improved through education, perhaps a carefully considered curriculum is the best medium to bring about this improvement.

Education, and specifically physical education, has reached a level of sophistication where serious thought must be given to a carefully reasoned and well-designed curriculum for the learner—one that can replace the disjointed divisions of the past. In fact, Birch (1992) explains that curriculum development can be critical to any school program. A well-planned, skillfully designed curriculum development process can produce not only meaningful content, but also program support from administrators, colleagues, and parents. A proper physical education curriculum must:

- Be conceived as an essential part of the total school curriculum
- Reflect the nature and needs of a democratic society in which respect for the interests and capacities of all individuals exists
- Be well balanced and afford varied experiences that will contribute to desirable outcomes for all age groups
- Be organized into well-planned experiences, extending from early childhood through post-secondary education and into later-life education
- Adhere to the philosophy, trends, methods, and materials of physical education in general
- Be based on rigorous criteria for content selection
- Be related to the health and guidance programs of the school
- Have an association with the community it serves

Building a curriculum from kindergarten to grade twelve is, as it should be, a challenging undertaking. It also provides an opportunity for professional excitement and expectation. On the other hand, if it is viewed as a red-tape procedure that is untimely, unwarranted, and painfully uninteresting, then the process is doomed to fail. Moreover, the people involved in its preparation will probably show little enthusiasm for its implementation.

The procedures used in planning a physical education curriculum will vary from one setting to another, but certain common concerns exist that need to be addressed. Tyler (1949) proposed four central questions that need to be addressed as the process of curriculum development begins.

1. What educational purposes should the school seek to attain?

2. What educational experiences can be provided that are likely to help attain these purposes?

3. How can these educational experiences be effectively organized?

4. How can people determine whether these purposes have been attained?

These questions provide a solid base from which the curriculum development process can begin. They can be restated in a four-step process as shown in figure 4.1.

Curriculum development is, fundamentally, a local responsibility. Schools that view it as such will be more successful because:

• A curriculum developed locally is more likely to meet the needs of its students.

• Curricular modifications and adjustments can be made more quickly on a local level.

• Personal involvement will increase the physical educator's commitment to the curriculum, its implementation, and evaluation.

Figure 4.1
Four-Step Process for Curriculum Development

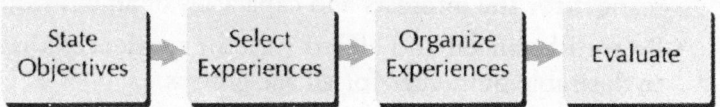

FACTORS AFFECTING CURRICULUM PLANNING

Thomas, Lee, and Thomas (2000) state that although the selection of a rational, theoretical model (that is, developmental, movement education, health-related fitness, etc.) is necessary for a foundation, other factors must be considered. These factors can be grouped into three categories as shown in figure 4.2.

Personal Factors

Personal factors deal with the learner. Since the program is ultimately designed for students, an understanding of their makeup is essential. The topic of pupil growth and developmental characteristics is especially germane to curriculum development. Selected activities need to coincide with the physical/physiological, intellectual, social, and emotional behavior of children at specific age levels.

Growth, as learning, is a continuous process. It is difficult to subdivide the growth period into specific age levels since children never abruptly complete one particular stage of development and begin the next. Moreover, a time never exists when

Figure 4.2
Factors Affecting Curriculum Planning

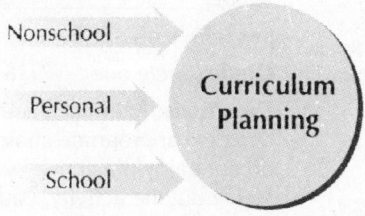

all children in a class are at the same growth stage. Chronological age and physiological age (maturation level) may be quite a distance apart. Chapters 7, 8, and 9 provide a detailed description of student characteristics and needs for the various age categories. In physical education, the level of physical maturity and development may have more to do with determining what to teach than it does in any other subject matter.

How students perceive physical activity and articulate the depth of their interest in physical activity can help shape the curriculum at any level from kindergarten to senior high school. Bain's (1979) study of perceptions showed that the way an individual perceives fitness, aesthetic qualities, and social dynamics determines the value he or she places on a physical activity. The complexity of the activity and the student's level of maturation also are factors. Therefore, a teacher can frequently gain first-hand information from watching students at play, listening as they talk with each other, and discussing student interests with parents and other teachers.

Students' interests can be used as a basis for establishing curriculum content. Interest inventories are self-report instruments in which students indicate likes and dislikes for certain activities. An inventory instrument designed to measure the level of activity interest of secondary school students is shown in figure 4.3. Surveying students' interests as part of a personal-factors approach may lead to a more meaningful curriculum for students. Additional invaluable information may be gained by adding open-ended questions. By asking students to indicate what they like and dislike about physical education, the strengths and weaknesses of the program can be ascertained.

One more source of information that can be used to guide curriculum planning is the student data collected on selected measures in the physical domain (see chapter 11). This data may include the student's medical status and such measures as posture and fitness assessment.

There has been a significant increase in student diversity in recent years. Before physical educators approach curriculum change, they need to accept that differences *do* exist. More importantly, educators need to

Figure 4.3
Activity Interest Inventory, Secondary School

Name: _____ **Date:** _____

Sex (circle one):   Female   Male

Grade (circle one):   7   8   9   10   **Age in years** (circle one):   13   14   15   16

Directions: After each activity listed below, indicate your interest in that activity by circling the appropriate choice. If you enjoy the activity you should circle "**L**" for LIKE; if you do not enjoy the activity you should circle "**D**" for DISLIKE; if you do not know how you feel about the activity, you should circle "**U**" for UNSURE.

Activity	LIKE	UNSURE	DISLIKE
Aerobic dance	L	U	D
Archery	L	U	D
Backpacking	L	U	D
Badminton	L	U	D
Basketball	L	U	D
Bicycling	L	U	D
Bowling	L	U	D
Conditioning	L	U	D
Fencing	L	U	D
Field hockey	L	U	D
Flag football	L	U	D
Floor hockey	L	U	D
Folk dance	L	U	D
Frisbee golf	L	U	D
Golf	L	U	D
Gymnastics	L	U	D
Handball	L	U	D
Hiking	L	U	D
Ice skating	L	U	D
Lacrosse	L	U	D
Martial arts	L	U	D
Modern dance	L	U	D
Orienteering	L	U	D
Pilates	L	U	D
Roller skating	L	U	D
Soccer	L	U	D
Softball	L	U	D
Square dance	L	U	D
Swimming	L	U	D
Table tennis	L	U	D
Tennis	L	U	D
Track	L	U	D
Volleyball	L	U	D
Wall climbing	L	U	D
Water aerobics	L	U	D
Wrestling	L	U	D
Other (list):	L	U	D
Other (list):	L	U	D

understand these differences and the implications they present for the curriculum. Jewett, Bain, and Ennis (1995) indicate that categories of student diversity include race, gender, social class, and style of learning. These authors provide both a complete discussion of each category and its effect on the learning of motor skills.

McCarthy (1990) warns that issues of diversity are generally looked at independently when, often times, a combination of two or more of these categories not only hinders a student's learning but also adversely impacts his or her opportunity to learn. The following example illustrates this situation. Diane was a ninth-grade student with a strong desire to participate in high school football. After making this desire known she had to endure ridicule, because of gender, not only from her peers but also from the coach, who expressed an unwillingness to work with her. In addition, she could not afford the participation fee. As a result, Diane experienced a double exclusion due to both gender and social class.

Too often, physical educators do not view students as an important factor in curriculum planning. The previous discussion makes it apparent that information about and from students is invaluable for this purpose.

School Factors

The school, as a social institution, is the primary agency responsible for the education of today's children. As a result, its influence on curriculum planning is extremely important. The selection of activities to include in a physical education curriculum is affected by a variety of school factors. These factors are shown in table 4.1. These factors cover a wide range of variables, from policy making at the administrative level to the availability and condition of facilities.

Table 4.1
School Factors Affecting Curriculum Planning

- Geographic location
- Facilities
  Indoor
  Outdoor
- Equipment
  Commercial
  Homemade
- Budget allocation
- Time allotment
  Days per week
  Minutes per day

- Faculty
  Gender
  Expertise
- Class size
- Class composition
  Coeducational
  Non-coeducational
  Students with disabilities
- Administrative policy
  School board
  Departmental

Nonschool Factors

Nonschool factors include items such as state and federal policy, evidence of contemporary curricular trends, and the community. A careful analysis of the values and beliefs of the community can prove beneficial (see table 4.2). What a community considers important and the attitudes of parents toward

---

Table 4.2
Community Analysis

What are the physical characteristics of the community?
- Find a map of the community and mark the boundaries of the area from which clients are drawn. Identify areas that are primarily residential or commercial.
- Are the residential areas single-family or multiple-family dwellings?
- Do residents rent or own?
- On the map mark the location of schools, parks, and recreation facilities.
- How readily can residents get to and from these facilities?
- Is public transportation available?

Who lives in the community?
- What is the age distribution of the residents?
- What is the ethnic and racial heritage of the residents?
- What languages are spoken?
- Describe the educational background and religious affiliations of the residents.

What is the economic base of the community?
- Who are the major employers?
- What kinds of jobs do the residents have?
- What is the average income?
- Do most households have one or more than one person employed?
- What kind of financial support is provided for public schools?

What is the political system in the community?
- What is the form of local government?
- Describe the political voting patterns of the area.
- Do residents belong to community action groups?

What is the culture in the community?
- What are the popular recreational activities of residents?
- What things are the residents proud of?
- What do the residents disapprove of?
- What do the residents hope for?
- How much crime occurs?
- How are outsiders viewed by residents?

From Jewett, A. E., and Bain, L. I. (1985). *The curriculum process in physical education.* Copyright © 1985 Wm. C. Brown Communications, Dubuque, IA. Used with permission of The McGraw-Hill Companies.

physical education have a definite bearing on curriculum development. Moreover, the concerns of both parents and community leaders cannot be underestimated as educators attempt to determine what to teach. Today, a community's educational priorities are being decided by parents and politicians. This fact alone indicates the need for public relations (PR), making it a professional responsibility. Baker (2001) tells us that educational programs often live or die as a result of public opinion, making high quality PR the *key* to maintaining public support. Schultz (1996) states that *PR = performance + recognition*. He further explains that for PR to be successful, schools must begin with a physical education program worthy of promoting. With that in place Baker (2001) lists the following strategies for gaining *recognition:*

- Promote locally. It is best not to rely on national efforts, but to remember that the most successful PR is one customized for a particular community.

- Publish a newsletter. Whether done once a month, once a semester, or once a year, a physical education newsletter can inform and educate parents as to what the physical education program offers and why.

- Showcase at school-wide functions. For example, schedule time during a PTA/PTO meeting to present a program in which students perform selected activities from the curriculum.

- Provide services for the faculty and the community. For example, coordinate a weekly fitness program at the school.

- Use the news media. Examples include (1) crafting a news release for the local newspaper on a specific physical education issue and (2) securing media coverage (newspaper, radio, television) for a specific event (PTA/PTO presentation, field day, etc.).

The availability of community resources (facilities and personnel) may affect the scope of activities than can be offered and thereby should be considered when developing the physical education curriculum. Possible resources include *facilities* such as bowling lanes, swimming pools, and golf courses, which are often available at little or no expense. To better plan, educators should conduct a survey of community resources (see figure 4.4). *Personnel* refers to individuals available to complement the program by serving as guest speakers, paraprofessionals, or part-time instructors.

STATE AND FEDERAL LAWS REGARDING PHYSICAL EDUCATION

State Requirements

Laws made at the state level govern which subjects are taught and how frequently in American schools. As a result, the yearly and daily require-

Figure 4.4
Community Resources Inventory

Part I: Facilities

Evaluator: _____ Date: _____

Community: _____

Instructions: Place an "X" if the facility exists and may be available for physical education.

	Indoor	Outdoor
Archery range	_____	_____
Badminton courts	_____	_____
Bicycle paths	_____	_____
Bowling lanes	_____	_____
Camping area	_____	_____
Dance studio	_____	_____
Golf course	_____	_____
Frisbee golf course	_____	_____
Golf driving range	_____	_____
Handball/racquetball courts	_____	_____
Hiking trails	_____	_____
Horseback riding stable	_____	_____
Lake area	_____	_____
Miniature golf course	_____	_____
Orienteering course	_____	_____
Rifle range	_____	_____
Sailing marina	_____	_____
Skating—Ice rink	_____	_____
Skating—Roller rink	_____	_____
Ski—Cross country paths	_____	_____
Ski—Downhill slopes	_____	_____
Swimming pool	_____	_____
Tennis courts	_____	_____
Weight-training facility	_____	_____
Other: _____	_____	_____
Other: _____	_____	_____
Other: _____	_____	_____

(continued)

Part II: Detailed Facilities Analysis

Instructions: Complete each of the following statements.

Name of facility: _____

Address: _____

1. What activities can be taught in this facility? _____

2. How far is the facility from the school? _____

3. Owner: _____

   Telephone number: _____

   E-mail: _____

4. Number of people facility may accommodate: _____

5. Equipment available for use: _____

6. Cost to use the facility: _____

8. Days and times when facility may be used: _____

9. Safety concerns:_____
   _____

10. Does the owner have insurance coverage? _____ yes _____ no

Comment(s): _____

Source: From Fig. 2.3, pp. 17–18 from *The lifetime sports-oriented physical education program* by William F. Straub. Copyright © 1996 by Prentice-Hall, Inc. Reprinted by permission of Pearson Education, Inc.

ments for physical education differ markedly among the 50 states. The National Association for Sport and Physical Education (2002) reported that 10 years after the *Goals 2000* called for inclusion of physical education as an integral component of public education, Illinois remains the only state requiring daily physical education. Fifteen years after Congress passed *Resolution 97* encouraging state and local governing bodies to provide *quality* physical education programs, only four states—Delaware, Illinois, Michigan, and Missouri—require elementary physical education specialists to teach elementary school physical education. Additionally, more than half of the states either have no requirements for physical education or require only one year in grades 9–12.

Title IX

Title IX of the Education Amendments was passed by Congress in June 1972. The provision requires that no person, on the basis of gender, be excluded from participation in, be denied the benefits of, or be subjected to discrimination under any program or activity receiving federal funding. Title IX has had a profound, national impact because nearly every school in the United States receives some form of federal financial assistance. By law all U.S. schools were required to complete a self-study during 1975–1976 to identify areas of noncompliance and develop strategies for full compliance by 1978. The result of this mandated self-study has been the identification and correction of various inequities. Title IX has no mandate for specific curriculum content, but states that equal access to the physical education curriculum must be maintained for boys and girls.

In addition to these legal implications, Siedentop (2004) indicates several practical implications of Title IX. These include the provision of coeducational classes, the assignment of physical educators according to expertise rather than by gender, and the grouping of students by ability rather than gender. It is perhaps not surprising, but certainly disappointing, that it took the weight of federal law to end gender discrimination in schools.

Public Law 94.142

In 1975 Congress passed the *Education for All Handicapped Children Act* (a more complete discussion of this topic appears in chapter 9). Its passage ensures a free, appropriate public education, including physical education, for all exceptional children. If at all possible, it is recommended that students be educated in the regular physical education program. However, specialized programming must be provided if necessary.

Physical education must be made available to all children.

SELECTION AND BALANCE IN PROGRAM PLANNING

Contributing to the lack of agreement on what to include in a physical education curriculum is the obvious fact that far more activities exist than can ever be implemented. As a person scans the extensive list of activities (see figure 4.3), from aerobic dance to wrestling, it becomes obvious that only those activities most appropriate for a given school can be considered.

Determining which activities to offer and how long to teach each activity are real challenges. "Six weeks for volleyball" is how one course of study is stated. "Eight weeks of gymnastics" is how another reads. How was this determined? The following issues have bearing on the selection and duration of program activities.

- The sequence of physical education activities in a school year
- The sequence of physical education activities over a span of several years
- Problems arising from a multi-activity curriculum model that attempts to do a little bit of everything all in one year
- The question of a repetitious curriculum that continues from year to year without change
- The relationship of the curriculum parts to the whole; for example, how much teaching of game skills should be conducted as physical education class work and how much should be covered in the intramural program
- The resources needed to provide a broad and varied curriculum

The issue of balance is always a basic consideration. Achieving harmony in physical education programming is not easily accomplished, particularly today when programs are often adopted or discarded due to political pressures that favor one approach over another, such as the push for activities to combat the obesity problem.

A BROAD AND VARIED CURRICULUM

The development of a broad and varied physical education curriculum, although theoretically sound, is not an easy task. When initiated, certain questions arise:

- How can the school provide a balanced program of physical activity amid various pressures for specialization?
- How comprehensive can a program of physical education be when many teachers have one or two favorite activities they like to stress, frequently at the expense of other meaningful, high-interest activities?

• How often have students said:

All we do is play basketball. Can't we be taught something else?

All we do is exercise. Why can't we do more activity?

All our physical educator does is teach soccer all fall. I wish we could learn something new.

Too often a physical education curriculum is thrown together piecemeal. This problem is compounded for many schools in which elementary and secondary school districts are organized and administered separately.

In extolling the value of physical education, many professionals have indicated that only a broad and varied program will appeal to all students. Programs are adequate when the various interests, needs, and abilities of both boys and girls are considered. This suggests a need for variety, including individual and self-testing activities, dual activities, rhythmics, and team sports that provide both vigorous and challenging experiences.

A broad and varied curriculum can be developed when physical education personnel and administrators see the need. Too often a curriculum lacks breadth because the majority of the budget, staff time, and even class time are devoted to interscholastic sports. Physical education programs will continue to suffer as long as athletics are the primary concern of the school.

STRUCTURING FOR QUALITY

Although this book is primarily concerned with activities and their organization in the physical education curriculum, it is almost impossible to avoid teaching methodology. Ultimately, the wisdom in choosing an activity and the talent in teaching that activity cannot be separated. Keeping this in mind is essential when a quality curriculum is the goal.

Being physically educated means the ability to use the body efficiently. However, too often the degree of recreation engaged in is directly proportional to the level of physical skill possessed for a given activity. This becomes obvious when you consider that Americans are not exclusively sports spectators but first-degree participants when they have the abilities. Furthermore, knowledge and ability in a sport produce the best kind of spectators—bright, sophisticated observers who express judgment as a result of experiences. Participants and spectators alike seek excellence. Both are interested not only in scores, reputations, and action but also in quality of performance.

Yet, because of poorly constructed curricula and ineffective teaching, thousands of schoolchildren have missed learning the fundamental movements and basic game skills at an adequate level of competency. The exposure to these skills has been minimal, at best. The actual learning period has been so minimized that only the idea of the activity has been taught, with little opportunity to put muscles through their paces. Student learners need to practice physical skills and slowly perfect them; but this concept has been treated too lightly. When physical education skills are properly learned, the lasting value takes the form of a solid kinesthetic experience, the kind of learning that provides the student with a deep appreciation for the skills. It is this appreciation that fuels the desire to engage in the activity again and again, perhaps throughout a lifetime.

SUMMARY

1. In this time of accountability, a thoroughly planned and well-designed physical education curriculum is a necessity.

2. Factors affecting curriculum development fall into one of three categories—personal, school, and nonschool.

3. Because of the diversity among students, an understanding of their growth changes, characteristics, needs, and interests is essential when developing a physical education curriculum.

4. School factors affecting curriculum development include budget, facilities, time allotments, teaching faculty, class size, class composition, and administrative policy.

5. Nonschool factors affecting curriculum development include state and federal policy/law, contemporary trends, and community resources.

6. Title IX has increased the sport and physical education opportunities for girls and women.

7. Public Law 94.142 mandated the provision of physical education opportunities for all exceptional children.

8. To best meet the varied needs and interests of a diverse student body, a broad and varied program is recommended.

QUESTIONS AND LEARNING ACTIVITIES

1. In the last several years, curriculum changes have been more widespread and intensive than at any time in the history of U.S. schools. Are these changes in the curriculum really significant? Do children learn better in the new programs than they did in the old ones? Have curriculum modifications been made simply because it was popular to do so?

2. Can a strong case be made for a national curriculum of physical education? What are the strengths and weaknesses of such a curriculum? Would you follow such a model if you had one?

3. React to the following statements:

   a. Planning the physical education curriculum *with* the instructors is far superior to planning *for* them.

b. Planning is essentially an administrative function calling for energetic leadership.

4. *Education* is a means of improving society and the *curriculum* is the medium to bring about such improvement. Do you agree or disagree with this statement? Justify your decision.

5. Answer Tyler's four central questions for curriculum development in each of the following situations:

a. a contemporary, large, multicultural, inner-city school

b. a small, homogenous, rural school

6. It has been said that the physical education teacher is the *key* to curriculum development. Explain this view.

7. What specific community characteristics may have an impact on the curriculum development process in your home town?

8. Explain how you might go about amending a secondary school physical education curriculum that was heavily game-centered.

9. Consider the meaning of the word *innovation*. Read what some general educators have had to say about it, discuss it with your classmates, and then respond to the following questions:

a. Is movement education a form of innovation in physical education?

b. Is substituting lacrosse for spring football practice an example of program innovation?

c. Are there examples of innovation you can suggest that may be appropriate for junior high school?

REFERENCES

Bain, L. (1979). Perceived characteristics of selected movement activities. *Research Quarterly, 50,* 565–573.

Baker, K. (2001). Promoting your physical education program. *Journal of Physical Education, Recreation, and Dance, 72*(2), 37–40.

Birch, D. (1992). Improving leadership skills in curriculum development. *Journal of School Health, 62*(1), 27–28.

Hass, G. (1987). *Curriculum planning: A new approach.* Boston: Allyn & Bacon.

Jewett, A., & Bain, L. (1985). *The curriculum process in physical education.* Dubuque, IA: Brown.

Jewett, A., Bain, L., & Ennis, C. (1995). *The curriculum process in physical education.* Boston: McGraw-Hill.

Kliebaard, H. (1968). The curriculum field in retrospect. In Paul W. F. Witt (Ed.), *Technology and the curriculum.* New York: Teachers College Press.

McCarthy, C. (1990). Race and education in the United States: The multicultural solution. *Interchange, 21,* 45–55.

National Association for Sport and Physical Education. (2002). *2001 shape of the nation report: Status of physical education in the USA.* Reston, VA: AAHPERD.

Schultz, B. (1996). Media mania: Ten steps to creating your own positive news media blitz. *Thrust for Educational Leadership, 26,* November/December, 10–12.

Siedentop, D. (2004). *Introduction to physical education, fitness, and sport.* New York: McGraw-Hill.

Straub, W. (1996). *The lifetime sports-oriented physical education program.* Englewood Cliffs, NJ: Prentice Hall.

Thomas, K., Lee, A., & Thomas, J. (2000). *Physical education for children: Concepts into practice.* Champaign, IL: Human Kinetics.

Tyler, R. (1949). *Basic principles of curriculum and instruction.* Chicago: University of Chicago Press.

Williams, T., & Underwood, J. (1969). *My turn at bat: The story of my life.* New York: Simon & Schuster.

5

RESEARCH AND CURRICULUM CHANGE

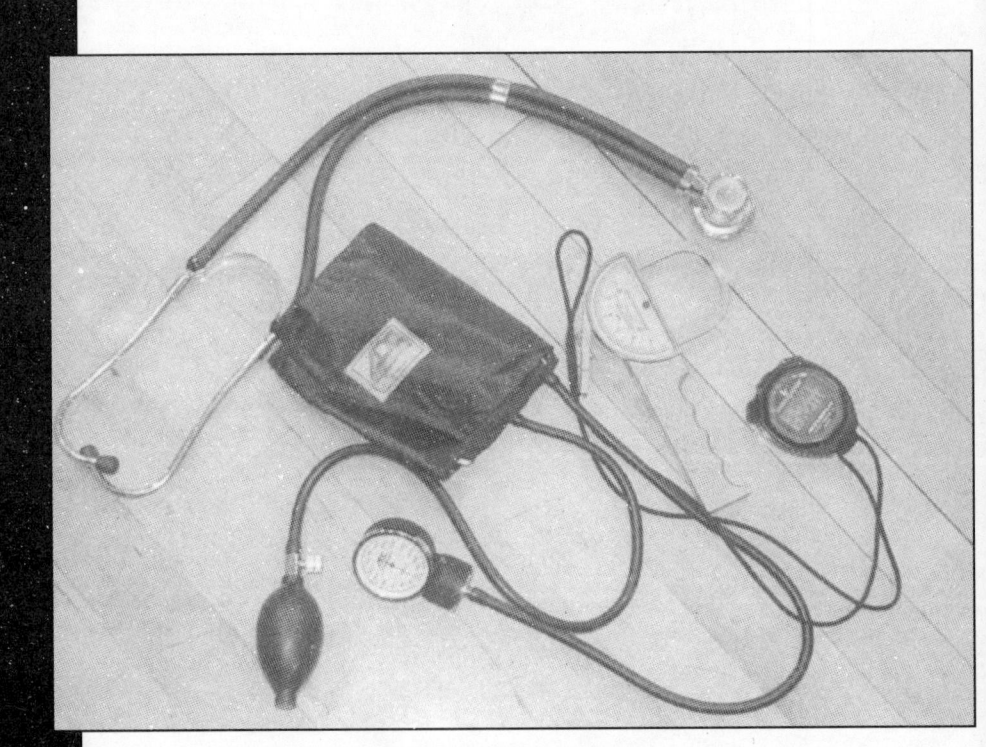

Outcomes

After reading and studying this chapter, you should be able to:

- Define
  - *Curriculum reform*
  - *Curriculum research*
  - *Reform*
  - *Research*
- Discuss AAHPERD's purpose and identify its five member associations.
- Discuss how a physical educator's resistance to change can affect curriculum reform.
- Discuss how a physical educator's eagerness for change can affect curriculum reform.
- Explain why professionals are reluctant to become involved in curriculum research.
- Identify a variety of professional periodicals and journals.
- Identify a variety of professional organizations.
- Analyze selected research studies to determine how they may impact curriculum change.

Everything is a result of change. This statement embodies an essential consideration that is the basis for curriculum development in physical education. In other words, change is inevitable and to face the educational implications arising from change is to also change.

Change is undoubtedly necessary; as society changes so do students and their needs and interests. In order for today's schools and the curricula within them to survive, especially in times of diminishing resources and overt criticism, these schools and their curricula must undergo change. This precept embodies the essential consideration underlying curriculum development, revision, and evaluation.

CHANGE AND CURRICULUM REFORM

Two concerns need to be discussed when facing change. The first concern is the educator's *resistance* to change; and the second factor is the educator's *eagerness* to change.

Why don't we want to change? Shirley Jackson's short story "The Lottery" portrays villagers who hold a lottery each year to decide whom they will stone to death. One character asks the villagers why they continue to carry out this inhumane ritual year after year. An elder quiets this inquisitive villager by answering, "Because we have always had a lottery." For the same reason that Tevya continued to "fiddle on the roof," a tradition was being upheld.

This also is the case with a variety of physical education practices, including inappropriate curriculum content, ineffective teaching methodology, and invalid evaluation procedures. Too many schools not only practice these traditions but also fail to seriously question them. Under such circumstances little effort and thought has been directed toward change. This is partly true because the process of change, creative renewal, and curriculum reform implies a break from the traditional viewpoint of "as it has been" to "as it ought to be." To initiate this renewal process requires energy and professional dedication because of the inherent resistance to change. Watson (1972) noted that all of the forces contributing to stability in an individual's personality and, in fact, in a social system can be perceived as a resistance to change. From a broader perspective, the tendencies of individuals and societies to resist change and preserve long-held practices, to "fiddle on the roof," is what ultimately threatens both individuals and the society in which they live.

The second concern, the eagerness to change, is characterized by overenthusiastic adaptability; in other words, "change for change's sake." As previously mentioned, change is inevitable. Many educators, however, equate change with progress. The result of this faulty logic is that educators often feel progressive and forward-thinking whether or not the change they embrace is warranted.

To combat this problem, educators must remember that curriculum reform should be based on some indication of need. To change just for the sake of changing is a waste of time and energy. Therefore, a basis for change is necessary. This basis may emerge from findings completed elsewhere or from local research. The tendency for people to sometimes move too fast to abolish present practices or to adopt new ideas may produce a change that, in retrospect, is too radical.

The process of curriculum reform is an ongoing one, responsive to political, economic, and other factors in the society at large. As our population continues to grow, new school districts are aligned, new facilities are constructed, and new physical education curricula are developed. In addition, existing physical education curricula are revised and updated.

Curriculum reform in the United States following World War II would have probably evolved much more slowly if it had not been for a prosperous, growing middle class comprised of ambitious parents who had great expectations for their children and envisioned education as the means of achieving them. This *baby boomer* group developed new communities, built new schools, and revitalized existing ones. Educators accepted this spirit of change and openly instituted school curriculum reform. Meanwhile, such influential factors as the population explosion, dramatic improvements in technology, and the proliferation of media outlets made curriculum developers more accountable for curricula that were both educationally sound and defensible. This era also witnessed an increase in private foundations whose missions included education reform and research, as well as an expanding role for the federal government in our nation's schools.

CURRICULUM RESEARCH IN PHYSICAL EDUCATION

Silverman and Ennis (2003) indicate that research is the basis for much of the decision making in our lives. Yet, surprisingly, research in physical education curricula has been somewhat limited. On the other hand, research on *teaching* physical education has been prolific. While this methodology research is obviously necessary since a teacher's effectiveness largely determines the degree of student learning, Anderson (1989) states that the curriculum in which the teaching occurs deserves more attention. In fact, it is the curriculum that provides both the context for teaching and the structure for student learning. Anderson poses some fundamental questions confronting physical educators:

- How are contemporary physical education curricula developed?
- How should physical education curricula be developed?
- What is occurring in today's physical education curricula?
- How effective are today's physical education curricula?

- What collaborative efforts between researchers and practitioners seem appropriate?

Siedentop (2004) adds these questions:

- What is the ideal physical education curriculum?
- Why do students voluntarily take part in physical education?
- To what extent do physical educators achieve their curricular objectives?
- What activities should be included in a contemporary elementary, middle school, or high school physical education curriculum?

Answering these questions is not an easy task, which might explain the dearth of research. Anderson (1989) identifies some of the inherent difficulties in conducting curriculum research. First, a physical education curriculum is an exceedingly complicated body for inquiry. Consider, for example, one physical educator teaching 40 students. The number of events occurring in a one-semester program is difficult to quantify, much less record and analyze. This complexity reveals limitations not present in controllable, laboratory research. The second reason educators are reluctant to undertake curriculum research pertains to the immense size of the physical education curriculum. Due to size, most research studies become intensive in nature because researchers are limited to a few programs in the overall curriculum or even a few cases in a single program. The third factor that contributes to the lack of research is the diversity of the physical education curriculum. This diverse nature presents a need for multiple data-collection techniques, including event recording, interview observation, and document analysis. Therefore, the results are multifaceted and have the potential for differing interpretations.

Regardless of whether an educator accepts the importance of curriculum research in physical education, it often becomes impractical due to lack of time, funds, and/or expertise. Fortunately, educators who are motivated to learn more can review existing research in a variety of professional periodicals and journals (see table 5.1). Despite this impressive array of publications, there is a consensus that today's physical educators tend not to use research in their work. In response, Lawson (1992) asks:

- Why should research be supported and completed if it is not used for human betterment?
- Why locate professional education programs in colleges and universities if practitioners do not use the theory and research?
- If practitioners tend not to use theory and research in their work, on what basis do they make decisions about their work practices?
- How do practitioners know whether they are helping or harming people?
- Is it possible for practitioners who do not use research to become obsolete?

Table 5.1
Professional Periodicals/Journals

ACSM's Health and Fitness Journal

Adapted Physical Activity Quarterly

American Health

American Journal of Health Education

American Journal of Health Promotion

American Journal of Health Studies

American Journal of Sports Medicine

American Physical Education Review

Athletic Administration

Athletic Therapy Today

Athletic Training

Australian Journal of Sports Medicine

Australian Journal of Sports Sciences

Baseball Research Journal

British Journal of Physical Education

British Journal of Sports Medicine

Canadian Journal of Applied Physiology

Canadian Journal of History of Sport

Canadian Journal of Sport Sciences

Clinical Journal of Sport Medicine

Clinical Kinesiology

Completed Research in HPERD

Corporate Fitness

Early Childhood Research Quarterly

European Journal of Sport Science

Exercise and Sport Sciences Reviews

Family Health/Today's Health

Health

Health Education

International Journal of Physical Education

International Journal of Sport Medicine

International Journal of Sport Nutrition and
  Exercise Metabolism

International Journal of Sport Psychology

International Journal of Sport Sociology

International Review for the Sociology of Sport

International Sports Journal

Interscholastic Athletic Administration

Journal of Aging and Physical Activity

Journal of Applied Physiology

Journal of Applied Sport Psychology

Journal of Community Health

Journal of Health Education

Journal of Human Movement Studies

Journal of Motor Behavior

Journal of Philosophy of Sport

Journal of Physical Activity and Health

Journal of Physical Education

Journal of Physical Education and Program

Journal of Physical Education, Recreation, and
  Dance

Journal of School Health

Journal of Sport and Exercise Psychology

Journal of Sport Behavior

Journal of Sport History

Journal of Sport Literature

Journal of Sport Management

Journal of Sport Rehabilitation

Journal of Sport Sciences

Journal of Sports Medicine and Physical Fitness

Journal of Teaching in Physical Education

Journal of the Canadian Association for Health,
  Physical Education and Recreation

Journal of the International Council for Health,
  Physical Education, and Recreation

Journal of Youth and Adolescence

Medicine and Science in Sports and Exercise

Middle School Journal

Motor Control

National Strength and Conditioning Association
  Journal

NIRSA Journal

Palaestra

Pediatric Exercise Science

Perceptual and Motor Skills

Physical Educator

President's Council on Physical Fitness and
  Sport Research Digest

Quest

Research Quarterly for Exercise and Sport

Scholastic Coach

Sociology of Sport Journal

Sport History Review

Sport Science Review

Sport, Education, and Society

Sports Medicine Bulletin

Teaching Elementary Physical Education

Today's Education

Women's Sports and Fitness

Research Studies

Although much of the published research may not relate directly to a specific curriculum, it may provide implications for curriculum change. Consider the following research studies and how each may influence curriculum reform.

Bryan, C., Johnson, L., & Solomon, M. (2004). Relationship between fitness testing and children's physical activity. *Research Quarterly for Exercise and Sport Supplement*, March, A-62.

The purpose was to examine the relationships between self-reported levels of physical activity and a health-related fitness assessment. Subjects were 108 fourth and fifth graders enrolled in daily physical education. The FITNESSGRAM was used to assess fitness and the Physical Activity Questionnaire for Children (PAQ-C) was used to gather data regarding physical activity levels. The PAQ-C asks children to indicate how many times in the previous week they participated in physical activity. Aerobic fitness was the dependent variable and body composition, BMI, and PAQ-C were the predictor variables. It was found that PAQ-C (self-reported physical activity levels) accounted for a significant proportion of variance in the children's aerobic fitness.

Brock, S., & Rovegno, I. (2002). A qualitative analysis of the influence of status of sixth-grade students' experiences during a sport education unit. *Research Quarterly for Exercise and Sport Supplement*, March, A-60.

The purpose was to gain an in-depth understanding of students' experiences and social interactions during a sport education unit of soccer. The subjects were 10 sixth graders, who comprised one of six teams in a 26-day soccer unit. Data collection included questionnaires, videotaping, and teacher observation of all student interactions during the unit. Students were observed in the classroom and during sport activities outside of school as well. It was found that students' experiences and interactions were strongly affected by their status. Students of high status made all decisions, dominated team discussions, played more, and received more privileges and attention from the teacher. They frequently silenced low-status students, who were subservient.

Davol, L., & Chepyator-Thomson, R. (2002). Factors contributing to female students' apathetic behavior in secondary school physical education. *Research Quarterly for Exercise and Sport Supplement*, March, A-65.

Previous research suggests that female students' interest in physical education declines progressively over the years. This study was undertaken to investigate factors that contribute to this apathy. Twenty-five females enrolled in a coeducational physical education class served as subjects. Data collection included questionnaires, school documents, and observa-

Research on the various elements of a physical education program can provide the kind of valuable feedback that underpins meaningful curriculum reform.

tions. It was found that three themes accounted for the students' apathy: (1) the teacher's repetitive teaching style; (2) the loosely organized setting; and (3) the activities taught. More specifically, due to these factors, students displayed nonactive behaviors and expressed boredom.

Derry, J., & Phillips, A. (2004). Comparisons of selected student and teacher variables in all-girl and coeducational physical education environments. *Physical Educator*, Late Winter, *61*(1), 23–34.

The purpose was to investigate selected student and teacher variables and compare the differences between these variables for female students and female teachers in single-sex physical education class and a coeducational physical education class. Eighteen female physical educators and their classes were selected. Nine teachers had single-sex (girl) classes while the remaining nine had coeducational classes. Both teachers and students were videotaped during classes. Students were administered two questionnaires. Results showed that female students from the single-sex class received a more positive learning experience. These students had more engaged-learning time and more student-initiated interactions with the teacher.

Greenwood, M., Stillwell, J., & Byars, A. (2001). Activity preferences of middle school physical education students. *Physical Educator*, Late Winter, *58*(1), 26–29.

The purpose was to determine the physical education activity preferences of middle-school students. Seven hundred fifty-one students, grades 6, 7, and 8, were asked to complete an activity interest inventory that included a 23-item checklist with activities ranging from archery to wrestling. It was found that all students had a strong interest in basketball, bicycling, swimming, and volleyball. Relative to gender, middle-school boys showed a greater interest in archery, flag football, soccer, and wrestling, whereas middle-school girls favored gymnastics, roller-skating, softball, and tennis.

Ha, A., Johns, D., & Shiu, E. (2003). Students' preference in the design and implementation of the physical education curriculum. *Physical Educator*, Early Winter, 60(4), 194–207.

The purpose was to examine students' expectations regarding the secondary school physical education curriculum. A PE questionnaire was given to 5,283 students. Results indicated that (1) one-third would not elect to take PE; (2) students were highly critical of their current curriculum; and (3) 75 percent thought that their opinions should be considered when developing the curriculum. In addition, male students favored basketball, fitness activities, and soccer, whereas female students favored badminton, tennis, and creative dance.

Huang, M., Chou, C., & Ratliffe, T. (2002). Relationship of children's fitness, physical activity, and physical education. *Research Quarterly for Exercise and Sport Supplement*, March, A-70.

The purpose of this study was to investigate second-grade students' physical activity levels during physical education classes that focused on skill development and health-related fitness. The researchers found (1) students in the fitness classes had a higher level of moderate to vigorous physical activity (MVPA) than those in the skill classes; (2) a significant negative correlation between the children's MVPA levels and the amount of teacher instruction; and (3) no difference between genders and MVPA levels within classes.

Larson, A. (2004). Student perception of caring teaching in physical education. *Research Quarterly for Exercise and Sport Supplement*, March, A-70.

Many teachers enter the profession because they care for children and are thereby regarded as "good" teachers because they are viewed as caring. It has been found that when students perceive their teachers as caring, they enjoy school and are motivated to learn. The physical education environment gives teachers an opportunity to spend a large amount of class time interacting with their students. The quality of this time can positively or negatively impact the student's physical education experience. The purpose of this study was to examine students' perceptions of caring physical education instruction. A *critical incidence* form was distributed to 389 K–12 students. The forms were coded and grouped the caring behavior into three clusters: the teacher (1) recognized me, (2) helped me learn, and (3) respected me. The researcher summarized the results by stating that numerous opportunities exist for the physical educator to exhibit caring behavior, and that students are aware of and appreciate the teacher's attention.

Menear, K. (2004). Use of high and low outdoor adventure elements to improve in-school behaviors of at-risk youth. *Research Quarterly for Exercise and Sport Supplement*, March, A-110.

This study was conducted to examine the relationship between an adventure-based intervention and self-perceptions of at-risk youth. The sample included 12 subjects between the ages of 12–16 who were at risk for academic failure and/or chronic misbehavior. The students participated in seven, 6-hour days of adventure-based activities during a five-month period. Activities were led by a certified adventure-based coordinator, a physical educator, and a school counselor. Students kept daily journals, which included their responses to questions dealing with self-awareness, group interactions, and school behaviors. Following the intervention, students were found to have improved school behavior and improved academics.

Miller, A. (2002). Middle school children's activity levels, physical self-perceptions, and physical self-importance differences. *Research Quarterly for Exercise and Sport Supplement*, March, A-76.

Research suggests there is a relationship between the physical activity level of children and how they view themselves physically. This study was designed to determine if differences existed between a child's activity level and ratings of self-perception and self-importance. The sample included 224 seventh and eighth graders. Using the Previous Day Activity Profile, students were required to rate their level of activity four times at random during a prescribed three-month period. In addition, they were asked to self-score their physical self-perception and physical importance. Results revealed that the students differed in self-perceptions relative to activity level. The more active students had higher ratings, whereas the less active students had lower self-perceived ratings. The researcher concluded that the more active seventh- and eighth-grade students are, the higher their physical self-perception, indicating that physical activity is an excellent contributor to how students view themselves.

OUTSIDE INFLUENCES

A growing number of national and international associations, councils, and institutes have become instrumental in sponsoring the kind of educational research that leads to curriculum reform (see table 5.2). These organizations are concerned, in varying degrees, with the promotion and advancement of physical education and its allied fields. Many of these exist as professional units dedicated to improving physical education at all levels, ranging from preschool to the university setting. Perhaps the most influential organization dedicated to physical education improvements is the American Alliance for Health, Physical Education, Recreation, and Dance (AAHPERD).

Table 5.2
Professional Organizations

Amateur Athletic Union
American Academy of Physical Education
American Alliance for Health, Physical Education, Recreation, and Dance
American Association for Health Education
American Association for Lifelong Recreation, Physical Activity, and Fitness
American Association of School Administrators
American Camping Association
American College of Sports Medicine
American Council on International Sports
American Heart Association
American Medical Association
American Medical Society for Sports Medicine
American Public Health Association
American Red Cross
American School Health Association
American Society of Biomechanics
American Society of Exercise Physiologists
American Sports Education Institute
American Sports Medicine Institute
Association for Supervision and Curriculum Development
Association for the Advancement of Applied Sport Psychology
Athletic and Recreation Federation of College Women
Athletic Institute
British Association of Sport and Exercise Science
British Society of Sport History
Educational Sports Institute
International Association of Physical Education and Sports for Girls and Women
International Association of Physical Education in Higher Education
International Association of Sports Law
International Council for Health, Physical Education, and Recreation
International Federation of Sports Medicine
International Society for Comparative Physical Education and Sport

International Society for Sport Psychology
International Society for the History of Physical Education and Sport
International Society of Biomechanics
International Society of Biomechanics in Sports
International Sociology of Sport Association
National Association for Girls and Women in Sport
National Association for Physical Education in Higher Education
National Association for Sport and Physical Education
National Association for the Education of Young Children
National Association of Governor's Councils on Physical Fitness and Sports
National Association of Secondary School Principals
National Council of Athletic Training
National Dance Association
National Education Association
National Federation of State High School Athletic Associations
National Parent Teachers Association
National Recreation and Park Association
National Safety Council
North American Society for Psychology of Sport and Physical Activity
North American Society for Sport History
North American Society for the Sociology of Sport
President's Council on Physical Fitness and Sports
Society of State Directors of Health, Physical Education, and Recreation
Sport Literature Association
United States Office of Education
United States Public Health Service
Young Men's Christian Association
Young Men's Hebrew Association
Young Women's Christian Association

American Alliance for Health, Physical Education, Recreation and Dance

Founded in 1885, AAHPERD has been instrumental in advancing physical education through a variety of services, including periodicals, special publications, conferences, and research. This educational organization supports, encourages, and provides assistance to member groups and their personnel as they initiate, develop, and conduct programs in each of the allied fields.

The Alliance specifically seeks to:

- Develop and disseminate professional guidelines, standards, and ethics
- Enhance professional practice by providing opportunities for professional growth and development
- Advance the body of knowledge in the fields of study and in the professional practice of the fields by initiating, facilitating, and disseminating research
- Facilitate and nurture communication and activities with other associations and other related professional groups
- Serve as their own spokespersons
- Promote public understanding and improve government relations in their fields of study
- Engage in future planning
- Establish and fulfill other purposes that are consistent with the purposes of the Alliance

AAHPERD's five member organizations, through which it accomplishes many of its goals, include:

- American Association for Health Education (AAHE)
- American Association for Lifelong Recreation, Physical Activity, and Fitness (AALRPAF)
- National Association for Girls and Women in Sports (NAGWS)
- National Association for Sport and Physical Education (NASPE)
- National Dance Association (NDA)

The American Association for Health Education (AAHE) envisions health education as a dynamic, collaborative process directed at protecting, improving, and promoting the health of all people. It serves professionals in a variety of settings, including health care, community agencies, businesses, K–12 schools, and institutions of higher learning. AAHE works to meet this objective through a comprehensive approach that encourages, supports, and assists health promotion professionals through education and other systematic strategies.

The newest of the five Alliance organizations is the American Association for Lifelong Recreation, Physical Activity, and Fitness (AALRPAF). It resulted from a merger of two long-standing organizations, those being the American Association for Active Lifestyles and Fitness (AAALF) established in 1949 and the American Association for Leisure and Recreation (AALR) established in 1938. AALRPAF is dedicated to enhancing quality of life by promoting active lifestyles through meaningful recreation, physical activity, and fitness experiences throughout a lifetime.

The National Association for Girls and Women in Sport (NAGWS) was founded in 1879 and strives to be the nation's premiere organization dedicated to addressing issues and promoting opportunities for all girls and women in sport. Its mission is to develop and deliver equitable and quality sport opportunities for all girls and women. It does this through relevant research, advocacy, leadership development, educational strategies, and programming that promotes social justice and change.

AAHPERD's largest member association is the National Association for Sport and Physical Education (NASPE), with more than 20,000 members, including K–12 physical educators, college and university faculty, researchers, coaches, athletic directors, and trainers. It is the only national association devoted exclusively to improving the total sport and physical education experience in the United States for all children and youth. It is comprised of individuals engaged in the study of human movement and the delivery of sport and physical activity programs. It conducts research and provides education programs in areas such as sport psychology, sport sociology, philosophy, history, and curriculum development. NASPE further develops and distributes public information that explains the value of physical education programs.

The National Dance Association (NDA), which began in 1932, promotes arts education for all individuals, from preschoolers to older adults. Its mission is to increase knowledge, improve skills, and encourage sound professional practices in dance education while promoting and supporting creative and healthy lifestyles through high quality dance programs. NDA contributes to the development of sound philosophies and policies for dance as education through conferences, conventions, and publications. It serves the profession in academic, recreational, and professional preparation at the local, state, and national levels.

Much of the scientific support for selected practices and activities in physical education school curricula today stems from work related to the AAHPERD. This support includes curriculum items related to physical fitness, sports skills, and activities for adapted physical education programs. Moreover, its publications include *American Journal of Health Education*; *Journal of Physical Education, Recreation, and Dance*; *Leisure Today*; *Research Quarterly for Exercise and Sport*; and *Strategies*.

In addition to AAHPERD, a number of other energetic, national groups exist that have a research influence. One such group is the President's Council on Physical Fitness and Sports (PCPFS). Established in 1956 by the executive order of President Dwight Eisenhower, this group serves as an independent presidential advisory council whose purposes are to:

- Enlist support of both individual citizens and private enterprise to promote physical fitness
- Initiate programs that inform the public about the importance of exercise
- Strengthen federal services and programs relating to fitness and sports
- Encourage state and local governments to emphasize the importance of physical fitness
- Develop cooperative programs among professional societies that encourage sound physical fitness practices
- Assist educational agencies, national sports bodies, and business and industry in developing physical fitness programs
- Encourage research in the sport and fitness area

The American College of Sports Medicine (ACSM), with its more than 20,000 members, promotes research and disseminates information dealing with the benefits and effects of exercise and the prevention and treatment of injuries incurred in sport and fitness activities. It is the world's largest multidisciplinary sports medicine and exercise science organization and is dedicated to improving the quality of life of people around the world by finding methods that allow everyone to live a healthier, more productive life. The ACSM's primary research publication is *Medicine and Science in Sports and Exercise*.

The National Association for the Education of Young Children (NAEYC), founded in 1926, has more than 100,000 members and is the nation's largest and most influential organization of early childhood educators and others dedicated to improving the quality of programs for children from birth through third grade. Its primary focus is to improve professional practice and working conditions in early childhood education and to build public support for high quality programs. Members include individuals involved in or providing programs for young children, such as teachers, directors, professors, and researchers. The NAEYC publishes the *Early Childhood Research Quarterly* and *Young Children*, journals covering developments in the practice, research, and theory of early childhood education.

The effects that such national groups have on school curricula are frequently underestimated. Physical education gets a boost several times each year from such organizations as the American Heart Association and the American Medical Association. These organizations disseminate both research funds and state and national health legislation information to its members and the general public.

In addition to the organizations, institutions, and agencies listed in this chapter, one additional state group exists whose influence is measurable. Within a state's department of education is a director or supervisor of physical education who provides influential leadership regarding physical education issues. Most states have one person in the director or supervisor position while others have several individuals assigned to assist individual school districts. This assistance typically takes the form of in-service programs, workshops, and clinics that are planned, promoted, and presented. It is these state department directors of physical education who, perhaps, are most effective at the grassroots level, helping local physical educators improve their own programs.

LOCAL ASSESSMENT

A program is best improved when changes are promoted after careful examination of research findings. These changes can occur in program areas such as content selection and sequencing, teaching methodology, and leadership. Research findings become more meaningful if they are drawn from self-assessment rather than from the examination of published findings done elsewhere. Therefore, physical educators are encouraged to evaluate their programs. A number of assessment instruments are available, including AAHPERD's Assessment Guide for Secondary School Physical Education Programs, Bucher's Checklist of Selective Items for Evaluating a Physical Education Program, and Bucher's Checklist and Rating Scale for the Evaluation of the Physical Education Program. A complete description of these and other instruments is included in chapter 11.

SUMMARY

1. As educators it is important to remember not to resist change or be too eager to accept change.
2. In recent years there has been abundant research on teaching physical education, but little research on curriculum.
3. Curriculum research is important because it is the curriculum that provides the context for teaching and the structure for learning.
4. Even though published research may not relate directly to an individual's curriculum, it may provide implications for curriculum reform.

5. A variety of national associations, such as ACSM, AAHPERD, and NAEYC, affect curriculum reform.

6. AAHPERD, with its five member associations, has been supportive in promoting research activities.

QUESTIONS AND LEARNING ACTIVITIES

1. It is widely believed that teachers should read scholarly publications. However, too few physical educators read journals such as the *Journal of Teaching in Physical Education, The Physical Educator,* and *Research Quarterly for Exercise and Sport.* Interview several people who have been teaching physical education five years or more. How do they feel about scholarly journals and the reading of research in general? Compare your findings with others.

2. There seems to be a shortage of research pertaining specifically to physical education program modification and the corresponding affect it has on students. Do you think this statement is correct? Give your reasons.

3. Define *reform*. How broad in scope is it? What does reform really mean when it is applied to educational practice?

4. Describe a situation in which curriculum revision can be brought about without research.

5. Define *innovation*. What is its derivation? Try applying the word to a certain grade level of physical education. List two or three ideas, procedures, or practices that would be innovative in a K–12 public school physical education curriculum.

6. Locate three current research articles and discuss how each may influence curriculum reform.

REFERENCES

Anderson, W. (1989). Curriculum and program research in physical education: Selected approaches. *Journal of Teaching in Physical Education, 8,* 112–114.

Brock, S., & Rovegno, I. (2002). A qualitative analysis of the influence of status of sixth grade students' experiences during a sport education unit. *Research Quarterly for Exercise and Sport Supplement,* March, A-60.

Bryan, C., Johnson, L., & Solomon, M. (2004). Relationship between fitness testing and children's physical activity. *Research Quarterly for Exercise and Sport Supplement,* March, A-62.

Davol, L., & Chepyator-Thomson, R. (2002). Factors contributing to female students' apathetic behavior in secondary school physical education. *Research Quarterly for Exercise and Sport Supplement,* March, A-65.

Derry, J., & Phillips, A. (2004). Comparisons of selected student and teacher variables in all-girl and coeducational physical education environments. *Physical Educator,* Late Winter, *61*(1), 23–34.

Greenwood, M., Stillwell, J., & Byars, A. (2001). Activity preferences of middle school physical education students. *Physical Educator*, Late Winter, 58(1), 26–29.

Ha, A., Johns, D., & Shiu, E. (2003). Students' preference in the design and implementation of the physical education curriculum. *Physical Educator*, Early Winter, 60(4), 194–207.

Huang, M., Chou, C., & Ratliffe, T. (2002). Relationship of children's fitness, physical activity, and physical education. *Research Quarterly for Exercise and Sport Supplement*, March, A-70.

Larson, A. (2004). Student perception of caring teaching in physical education. *Research Quarterly for Exercise and Sport Supplement*, March, A-70.

Lawson, H. (1992). Why don't practitioners use research? Explanations and selected implications. *Journal of Physical Education, Recreation, and Dance*, 63(8), 36, 53–57.

Menear, K. (2004). Use of high and low outdoor adventure elements to improve in-school behaviors of at-risk youth. *Research Quarterly for Exercise and Sport Supplement*, March, A-110.

Miller, A. (2002). Middle school children's activity levels, physical self-perceptions, and physical self-importance differences. *Research Quarterly for Exercise and Sport Supplement*, March, A-76.

Siedentop, D. (2004). *Introduction to physical education, fitness, and sport*. Mountain View, CA: Mayfield.

Silverman, S., & Ennis, C. (2003). *Student learning in physical education*. Champaign, IL: Human Kinetics.

Watson, G. (1972). Resistance to change. In G. Zaltman, P. Kotler, & I. Kaufman (Ed.), *Creating social change*. New York: Holt, Rinehart & Winston.

6

PROGRAM ORGANIZATION

Outcomes

After reading and studying this chapter, you should be able to:
- Define
    - *Ability grouping*
    - *Computer-based instruction*
    - *Curriculum guide*
    - *Cycle plan*
    - *Paraprofessionals*
    - *Peer teachers*
    - *Performance contracts*
    - *Scope and sequence*
    - *Selective program*
    - *Student assistants*
    - *Team teaching*
- Identify and define criteria for organizing physical education activities.
- Distinguish between conventional and flexible scheduling.
- Describe the three approaches to team teaching.
- Describe the procedure for developing a curriculum guide.
- Identify advantages, disadvantages, and implications arising from the following innovations:
    - *Ability grouping*
    - *Computer-based instruction*
    - *Paraprofessionals*
    - *Peer teachers*
    - *Performance contracts*
    - *Student assistants*
    - *Team teaching*

Planning an effective instructional program requires the development and harmonious arrangement of curriculum content, teaching methodology, and evaluation practices. As this is accomplished, a real opportunity exists for a school system to correct past errors, to be innovative, and to reform. The extent to which a curriculum is reformed and, thereby, improved depends largely on how much interaction occurs between the individuals involved in the process. Just as the first law of ecology is interaction, so is interaction the first rule of successful educational organization. When these interactions include all individuals involved—administrators, teachers, parents, students, and concerned community members—there is no limit to what can be achieved. As these relationships develop, an understanding of the need for physical education occurs.

THE CURRICULUM GUIDE

It has been said that the world belongs to those people who know where they are going. It only follows that if such people know where they are going, they will certainly know if and when they get there. In education, however, it is not enough to know where you are going. In this age of accountability, such information must be made available for all concerned with the educational process, including administrators, parents, and students. Unfortunately, too many physical educators have known where they wanted to go but failed (1) to communicate this information effectively and (2) to document their intent as a matter of record. These lost opportunities highlight the need for writing a course of study, or curriculum guide. The curriculum guide is a framework that contains, among other things, the program's philosophy, general objectives, scope of offerings, and sequence in which these offerings will be presented. Its development is time consuming and often tedious. When completed, however, what emerges is an administratively sound document that gives direction to the physical education program. The curriculum guide contains *what* a physical education program purports to do and *how* it purports to do it. Regardless of the school system's size, a need exists to present the K–12 physical education curriculum on paper for everyone to see, review, and periodically revisit.

Annarino, Cowell, and Hazelton (1986) define a curriculum guide as a document for both teachers and students of physical education that indicates how educational philosophy and theory are translated into action. Rink (1993) states that the curriculum guide should specify what a student should learn and behaviorally be able to do after the educational program is completed. Melograno (1996) states that a curriculum guide should provide answers to the following three questions:

1. Where are we going? This question is best answered by establishing a sound philosophy and a list of general objectives for physical education.

2. How will we get there? This question is best answered by (1) planning and establishing student outcomes and explicit content, and (2) devising meaningful learning experiences.

3. How will we know we have arrived? This question is best answered by selecting and/or developing valid evaluation procedures for determining the effectiveness of the curriculum.

A useful curriculum guide requires forethought and planning, but even then it will have limitations. The guide alone cannot fix a weak program. The most important element of a quality program is instruction, and it is the teacher's responsibility to successfully provide this instruction. A well-developed curriculum guide, however, provides a strong beginning.

A theoretical organizational flow chart for curriculum development in physical education is illustrated in figure 6.1. This chart is most applicable for a large school district with at least one physical educator at each of the four subcommittee levels shown. A discussion of the process using this model follows; however, this model can and should be modified to fit within the structure of individual school districts.

Figure 6.1
Flow Chart for Curriculum Development in Physical Education

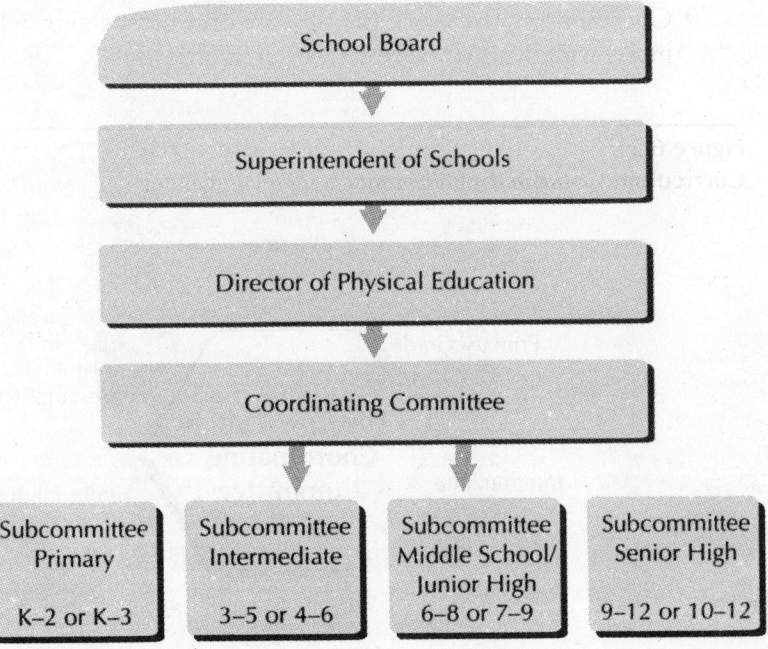

The Curriculum Coordinating Committee

The curriculum coordinating committee (CCC) is the group charged with writing the guide. This committee is comprised of 5 to 6 individuals acting as "coordinators" for the various subcommittees. The makeup of the CCC, shown in figure 6.2, should include one administrator, the physical education director, and one representative from each of the four levels of the physical education program. One of the primary tasks of the CCC is to explore the degree of faculty satisfaction with the present curriculum. A curriculum survey can be prepared so that all physical education faculty, fellow teachers, and/or students may respond. Another core task is to meet with individual school principals and eventually with the school board in order to establish necessary administrative procedures. These procedures may include (1) obtaining release time for committee members and (2) making budget provisions for the employment of a curriculum consultant if needed.

The CCC also is responsible for:

- Formulating the foundations on which the guide will be written, including the mission statement, the philosophy of physical education, a listing of the program's general objective, and information on the characteristics, needs, and interests of the community's youth
- Selecting faculty members for each subcommittee
- Formulating work schedules for all subcommittees
- Establishing procedures underlying the guide construction effort
- Coordinating the work of the subcommittees and, when completed, integrating it into the broader curriculum

---

Figure 6.2
Curriculum Coordinating Personnel

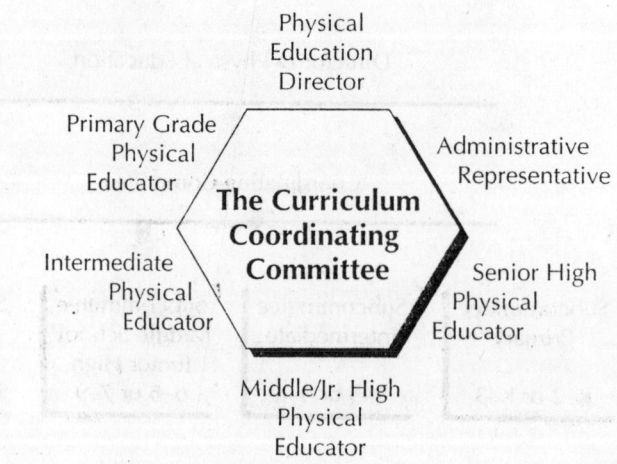

- Providing the subcommittees with information as needed
- Providing the subcommittees with regulations and/or requirements from the state department of education
- Collecting, reviewing, and collating all materials from the subcommittees
- Writing the curriculum guide

Curriculum Subcommittees

The subcommittees should be comprised of one physical educator, one principal, one nonphysical education teacher from the specific grade level, one parent, and one student (see figure 6.3). The regular classroom teacher can be an asset during the content selection and arrangement process because he or she views this content differently than the physical educator. Parents are generally helpful because they are keenly aware of the needs and interests of the children in the specific community. A concerned parent who is willing to work diligently is the ideal. However, the selection of this individual may cause problems if chosen arbitrarily. It is recommended that the responsibility for selecting which parents should serve on each of the four subcommittees be given to the school system's parent group, generally called either the Parent Teacher Association (PTA) or Parent Teacher Organization (PTO). At both the junior high and senior high school levels, a student should be added to the subcommittee. Each subcommittee should:

- Determine the most effective curriculum model for its grade level(s)
- Select and organize the physical education activities for its grade level(s)
- Provide a scope and sequence for these activities
- Establish student behavioral objectives or outcomes
- Select appropriate methods for the evaluation of students

The work of specific subcommittees is time consuming. The philosophical exchange, the debate over personal beliefs and convictions, and the arrival at behavioral outcomes for youth at various age levels cannot be accomplished in a matter of days or even weeks. In a cooperative effort, some common agreement and enthusiasm for the content of the guide is likely to occur. When teachers and parents work together to plan a physical education program, the result is not only a meaningful curriculum for the students but also professional growth for the teachers. In addition, parents gain a more complete understanding of the place and importance of physical education. In fact, the cooperative effort of developing a curriculum guide is capable of providing a rich professional experience for everyone involved, including administrators, teachers, parents, and students.

Proceeding slowly and carefully in the initial meetings is important in order to lay a foundation upon which physical education teachers can construct meaningful physical education experiences when the guide becomes a reality.

Figure 6.3
Subcommittee Personnel

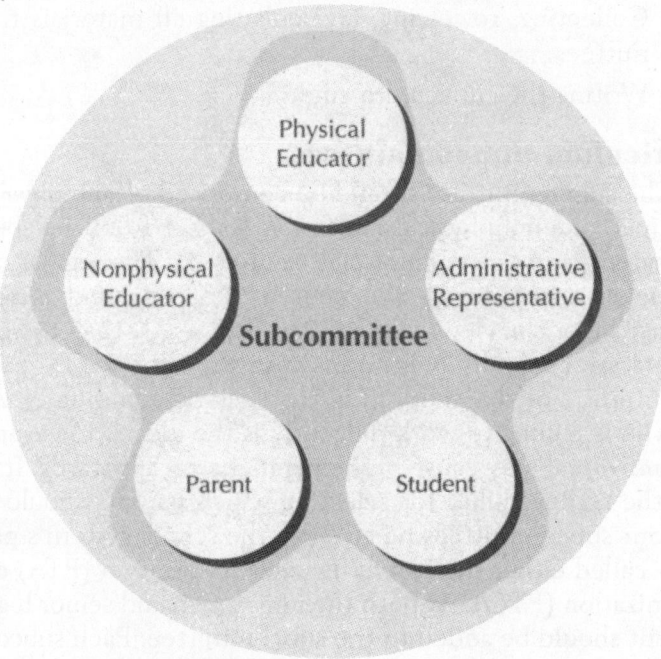

The Administrator's Role

The support of the school system's administration is crucial if the development of the curriculum guide is to be successful. This support can best be demonstrated by the principal serving on the curriculum coordinating committee. The administrator's specific responsibilities are to:

- Arrange meeting places for the CCC and the subcommittees
- Arrange release time, if necessary, for CCC and subcommittee members to meet
- Make necessary resources available, including instructional materials, printed references, and consultant services
- Arrange for necessary equipment, supplies, and secretarial services
- Cooperate with committee personnel in both raising questions and seeking answers relating to the impact of the new curriculum guide on the rest of the school curriculum
- Inform the school's governing body as to why both the guide and physical education are important for students and the community
- Inform the school's governing body as to the progress being made in developing the curriculum guide

Constructing the Guide

When the curriculum materials have been collected from the subcommittees, but before the writing begins, an examination of other curriculum guides is recommended. Libraries at most teacher education colleges and universities contain examples of curriculum guides from local school systems. Practical guides are often prepared at the state level. Greenwood and Stillwell (1999) have identified 42 state education agencies that have developed a recommended physical education curriculum guide, course of study, frameworks, or standards. Many of these can be obtained free of charge by writing to the respective state agency. Table 6.1 provides a sample

Table 6.1
State Agency Curriculum Materials

Title	Source
Arkansas Physical Education and Health Curriculum Frameworks (2002)	Arkansas Department of Education 4 Capitol Mall Little Rock, AR 72201 http://www.arkedu.state.ar.us/
California Physical Education Framework (2002)	California Department of Education 1430 N Street Sacramento, CA 95814 http://www.cde.ca.gov/
Louisiana Physical Education Content Standards (2003)	Louisiana Department of Education 1201 North 3rd Street Baton Rouge, LA 70804 http://www.doe.state.la.us/lde/index.html
North Carolina Healthful Living Standards Course of Study and Grade Competencies (1996)	North Carolina Department of Public Instruction 301 North Wilmington Street Raleigh, NC 27601 http://www.dpi.state.nc.us/
New Jersey Core Curriculum Content Standards for Comprehensive Health & Physical Education (2004)	New Jersey Department of Education P.O. Box 500 Trenton, NJ 08625 http://www.state.nj.us/education/
Pennsylvania Academic Standards for Health, Safety, & Physical Education (2002)	Pennsylvania Department of Education 333 Market Street Harrisburg, PA 17126 http://www.pde.state.pa.us/
South Dakota Physical Education Standards (2000)	South Dakota State Department of Education 700 North Illinois Street Pierre, SD 57501 http://www.state.sd.us/deca/
Physical Education Standards of Learning for Virginia Public Schools (2001)	Virginia Department of Education P.O. Box 2120 Richmond, VA 23218 http://www.pen.k12.va.us/

listing of state agency curriculum materials from eight states. For a comprehensive search the reader is directed to http://ideanet.doe.state.in.us/htmls/states.html, which is the Web site for state education departments.

Curriculum guides often contain consistent content but varied formats. Content is generally categorized by grade level. The most common format separates content into the following grade levels: elementary school, middle or junior high school, and senior high school. Some guides further subdivide the elementary school content into primary (K–2 or K–3) and intermediate (3–5 or 4–6). Regardless of the number of levels, the content must be coordinated to avoid undue repetition and to ensure that the activities are sequenced progressively. Of course, this is the task of the curriculum coordinating committee. If the guide is intended for the total physical education program, the format shown in table 6.2 may be employed for its construction.

In addition to these basic components, the following information may be included in a more specific curriculum guide:

- Developmental characteristics of students
- Organizational formats, both unit and lesson
- Definitions specific to physical education
- Instructional strategies
- Grading policy

Table 6.2
Physical Education Curriculum Guide Format

   I. Title
  II. Introduction
 III. Physical Education Philosophy
 IV. Justification and Need for Physical Education
  V. General Objectives
     A. Specific objectives
     B. Behavioral objectives by grade level
 VI. Program Content
     A. Scope and sequence
       1. Primary grade level
       2. Intermediate grade level
       3. Middle school level
       4. Senior high school level
     B. Time allotment
     C. Class size and composition
 VII. Evaluation in Physical Education
     A. Student
     B. Teacher
     C. Program
VIII. Resources and Instructional Materials
 IX. Selected References

- Accident policy
- A map (blueprint) of school, highlighting teaching stations

Jewett, Bain, and Ennis (1995) state that the contents and degree of specificity of the curriculum guide reflect the creators' philosophy about the students who will be taught and the teachers who will implement the curriculum. Guides that are more flexible and/or less detailed will allow teachers the autonomy to make decisions regarding content, methodology, grading, and so on. Guides that are precise in content and well detailed will provide teachers, especially the new ones, a blueprint from which to teach.

SCOPE AND SEQUENCE

Scope relates to the *breadth* or *quantity* of movement experiences in the physical education curriculum. It refers to *what* will be taught at all grade levels. As previously mentioned, the scope of a curriculum will vary depending on changes in society and the needs and interests of students. Sequence relates to the *quality* of movement experiences in the physical education curriculum. It refers to *when* activities should be taught, the order in which the content is to be delivered, and the grade placement of the physical education experiences. Sequence defines the curriculum horizontally, whereas the scope defines it vertically (see table 6.3).

Scope

In determining scope, it is a good practice to first view the entire physical education curriculum, K–12. This view tends to put content in perspective and gives curriculum planners a chance to consider a broad and varied curriculum, as discussed in chapter 4. Once the spectrum of activities is clearly visible, the next step is to break it down into categories of content appropriate for such organizational divisions as the elementary grades or senior high school grades. Content at these various levels will be discussed more thoroughly in chapters 7 and 8.

In the past, judgments relative to the K–12 scope of the physical education curriculum were frequently subjective and arbitrary. Too often curricula are developed in isolation, with each physical educator in a school developing a curriculum for his/her classes. This is yet another argument for developing a curriculum guide.

Bookwalter (1964) provides a valid set of selection criteria to use in dealing with the scope of a physical education program. Although they are more than forty years old, these criteria are as timely today as they were when first published. The criteria state that the activity selected must:

- Contribute to the general objectives of physical education

Table 6.3
Junior High School Scope and Sequence for Badminton, Golf, and Tennis

Activities	Grade 7	Grade 8	Grade 9
Badminton			
Nature of game	✓		
Court	✓		
Singles	✓		
Strategy	✓		
Rules	✓		
Doubles		✓	✓
Strategy		✓	✓
Rules		✓	✓
Alignment		✓	✓
Court etiquette	✓	✓	✓
Grip	✓	✓	✓
Forehand	✓	✓	✓
Backhand	✓	✓	✓
Stance	✓	✓	✓
Skills	✓	✓	✓
Overhead clear		✓	✓
Around the head clear			✓
Overhead drop			✓
Long serve	✓	✓	✓
Short serve	✓	✓	✓
Smash		✓	✓
Footwork	✓	✓	✓
Front court	✓	✓	✓
Back court	✓	✓	✓
Golf			
Nature of activity	✓	✓	✓
Equipment selection	✓	✓	✓
Care of equipment	✓	✓	✓
Rules	✓	✓	✓
Etiquette	✓	✓	✓
Grip	✓	✓	✓
Stance	✓	✓	✓

Note: When an item checked for grade 7 is also checked for grades 8 and 9, that skill is reviewed in the later grades.

(continued)

Activities	Grade 7	Grade 8	Grade 9
Swing	✓	✓	✓
Skills	✓	✓	✓
Putting	✓	✓	✓
Chipping		✓	✓
Short irons	✓	✓	✓
Long irons	✓	✓	✓
Woods	✓	✓	✓
Tennis			
Nature of activity	✓	✓	✓
Court	✓	✓	✓
Singles	✓	✓	✓
Strategy	✓	✓	✓
Rules	✓	✓	✓
Doubles		✓	✓
Strategy		✓	✓
Rules		✓	✓
Alignment		✓	✓
Court etiquette	✓	✓	✓
Grip	✓	✓	✓
Eastern	✓	✓	✓
Continental	✓	✓	✓
Stance	✓	✓	✓
Skills	✓	✓	✓
Forehand	✓	✓	✓
Backhand	✓	✓	✓
Serve	✓	✓	✓
Lob		✓	✓
Volley		✓	✓

- Be within the physical, physiological, and intellectual capacities of the learner
- Be meaningful and purposeful to the learner
- Offer frequency of participation outside of the school environment
- Have carry-over value and/or lead to further activity involvement
- Be feasible relative to the school's resources (time, facilities, staff, etc.)

Sequence

The physical education curriculum content needs to be organized so that students will be able to progress toward an increasingly mature utilization of both their knowledge and skills. This kind of development calls for a careful sequencing of activities. The criteria for guiding the organization of activities include the following.

Progression. Not all physical education activities are equal in complexity regarding the type (individual, team, etc.) or skills within an activity (tennis forehand, backhand, and serve). The form required, the competency level expected, and the practice needed to develop a skill are all functions of its relative difficulty. The activity may be so easy that it is boring or so difficult that it is frustrating or unsafe.

Variety. A broad choice of physical education activities is necessary to meet all of the objectives of a physical education curriculum. These activities can be classified into selected categories; for example, individual, team, aquatics, gymnastics, and rhythms. Allotting a proper amount of time for each category is necessary for a balanced program.

Seasonality. To heighten student interest, it is necessary to give attention to fall, winter, and spring activities. Planning a flag football unit of instruction for the fall, a basketball unit for the winter, and a golf unit for the spring will better fit the customs and expectations of society, thereby creating a more conducive environment for learning.

Feasibility. This is partly inherent and partly contrived. A school without a swimming pool will have a difficult time establishing an aquatics program. Possibilities are minimized when a 2-day-a-week program exists instead of a 5-day-a-week program. More can be taught with a 1-to-1 ratio of equipment-to-child than with a ratio of 1 to 10. Twenty-minute periods of physical education provide less time for movement experiences than 50-minute periods. Classes of 60 students will undoubtedly allow less time-on-task for individuals than a class of 25.

Practice for mastery. Allowing ample time to learn (master) a movement is a function of sound organization. An adequate amount of movement time (practice) is necessary for students to be physically educated. The goal is for students to overlearn the skills so that they become automatic. When students reach a level of competency in the skills they have learned, their enjoyment level will increase as well.

All five of these criteria warrant consideration when organizing content. There are too many instances of repetitious content where the same activities and the same skills within these activities are taught from year to year. Too often, there is not enough progression of content so that selected activities can be learned to a higher level of competency. Such progression occurs in mathematics where the student studies algebra, completes it, and goes on to geometry. But in physical education, students are instructed in volleyball every year from grades four to twelve, with little, if any, change in content.

One of the biggest obstacles to a quality program is when the curriculum fails to provide proper sequencing of learning experiences. If curriculum developers would take time to provide a well thought out sequence of learning activities for all of the physical education activities, K–12, the ultimate consequences could be spectacular. Table 6.4 presents an example of how the skills in a volleyball unit of instruction might be sequenced to ensure learning.

Table 6.4
Progression for Volleyball Content

Skills	Grade 6	Grade 7	Grade 8
Underhand serve	I	R	—
Overhand serve	—	I	R
Underhand set	I	R	R
Overhead set	I	R	R
Block	—	—	I
Spike	—	I	R
Dink	—	—	I

I = Introduce
R = Review to competency

Sequencing is just as important for content in the elementary school, particularly that which deals with the fundamental movement skills, as it is for the more advanced, sport-specific skills associated with (for example) learning to play basketball. Table 6.5 illustrates a sequence that ranges from easy to difficult tasks that can be organized for the fundamental locomotor skill of walking.

SCHEDULING THE CURRICULUM

Scheduling is an administrative procedure. No matter how ideal the physical education curriculum may appear on paper, its success will be determined by how well it is implemented. Its implementation involves scheduling to some degree and as such, prompts four questions:

- How much time should be set aside for physical education?
- How should this time be scheduled at each level (elementary school, middle school, junior high school, and senior high school)?

- Should physical education be scheduled two days a week, three days a week, or five days a week?
- Is the schedule flexible enough to vary according to both the nature of the activities and the availability of facilities?

Although computers have made scheduling an easier task, the fundamental decision of what goes into the computer is still the responsibility of the administrator.

There are two basic organizational patterns used for scheduling—the traditional pattern and the block pattern. The traditional schedule divides

---

Table 6.5
Handling the Body in Relation to Walking

1. Does the student
   - Know the difference between a relaxed and stiff walk?
   - Swing arms naturally with alternation of legs?
2. Adapt walking in relation to
   a. Sound
      - Can the student clap the even beat?
      - Can the student walk correctly to the clapping rhythm?
   b. Tempo
      - Can the student walk fast and slow?
      - Can the student walk with the beat using variations in tempo?
   c. Movement
      - Can the student walk while moving other body parts (hands, arms, etc.)?
      - Can the student walk using a zigzag route?
   d. A partner
      - Can the student walk beside a partner?
      - Can the student walk behind a partner?
   e. Music
      - Can the student walk in time to a song?
      - Can the student walk with others with various types of accompaniment?
   f. Space
      - Can the student walk freely among classmates?
      - Can the student walk as the play area decreases (modified by boundaries)?
   g. Imagery
      - Can the students walk as if they were on ice?
      - Can the students walk as if they were elephants?
3. Activities requiring walking
   - Partner tag
   - Posture Tag
   - Hokey Pokey

the school day into an equal number of class periods (7–9) that are all the same length (40–55 minutes). This pattern makes it easier to (1) keep class sizes relatively constant, (2) make classes heterogeneous in make-up, and (3) organize for coeducational experiences, if desired.

The second scheduling approach is the block pattern. Rettig and Canady (1999) feel that block scheduling provides a more efficient way to organize the school day. It allows students to spend more time (80–100 minutes) in fewer courses per day. Block scheduling usually takes one of two forms. With the first form, all students take 4–5 courses that meet 5 days a week for the same amount of time. The second form allows students to take 3–4 courses on Mondays, Wednesdays, and Fridays; and 2–3 different courses on Tuesdays and Thursdays. Block scheduling for physical education has several advantages: (1) both students and teachers have fewer preparations per day, (2) instructional time is increased, as less time is spent transitioning from class to class, (3) the lengthened class periods allow for nontraditional activities that may require travel (golf, bowling, roller skating, etc.), and (4) discipline problems are reduced (Bukowski & Stinson, 2000).

In most schools the traditional approach to scheduling is still practiced, and it is easy to understand why school administrators are reluctant to change to a flexible scheduling of classes. It is much simpler to divide the school day into six, seven, or eight standard-length periods and assign the same amount of time for all subject matter. The assumption now, however, is that a higher quality of physical education can be provided when more effective schemes of organizing school time are used.

The Cycle Plan

As previously mentioned, the tendency in physical education, especially at the secondary level, is to teach the same major activities each year. Sometimes repetition is good, particularly if it reacquaints students with topics or activities they may have missed the previous year. However, it may be more effective in terms of both student motivation and ultimate retention to teach certain seasonal activities every other year. This type of arrangement is called the *cycle plan*.

When the total physical education program has been carefully planned, the value of the cycle plan may emerge. For example, instead of trying to teach short units of several fall, winter, and spring sports each year, effort could be concentrated on only half of these one year and on the other half the following year. This has practical application especially in climates where the outdoor season is short. This arrangement provides for more intensive study by allotting more time for each activity. Of course, the implication arising from this approach is that students must be enrolled in physical education for two successive years in order to truly benefit from all activities.

TEACHING STATIONS

Teaching stations are defined as any area used for the instruction of physical education. The more traditional teaching stations for both indoor and outdoor use are shown below.

Indoor	Outdoor
Classroom	Football field
Dance studio	Soccer field
Gymnasium	Softball field
Swimming pool	Tennis courts
Weight room	Track

Evidence exists that supports the thesis that the finest curricula, in both content and instruction, occur where both indoor and outdoor multiple teaching stations are present. Therefore, the current increase in campus-type schools, consolidated schools, and centralized schools has much to offer physical education. In some instances, more than 25 percent of the indoor space is assigned to physical education. With vastly improved outdoor playing fields and courts, it becomes easier to provide and schedule large-group, small-group, and individual activities.

The majority of elementary school buildings constructed in the last decade have bright, optimum-sized gymnasia and playing fields in addition to traditional playgrounds. Secondary schools have auxiliary gymnasia, separate weight-training facilities, and dance studios. Increasingly, the most promising facility being constructed in the United States is the field house, with its wide-open spaces, multi-use fixtures, and portable floors. The field house sets the stage for a more comprehensive physical education curriculum. With increased instructional staff assigned to large groups of students and the excellent variety of field house teaching stations, there is the potential for conducting a superior program.

TEAM TEACHING

The essential requirement for team teaching is that two or more teachers be scheduled in a way that enables them to work together to carry out the program more efficiently than if they were working by themselves. Team teaching is a way to better meet the needs of both boys and girls in coed classes by having both a male and a female teacher share the responsibilities of planning, presenting, and evaluating. This is not a new concept but is a practice that has potential when appropriately used.

Three approaches to team teaching can be employed (see figure 6.4). The *unit team approach* is somewhat hierarchical. One teacher is desig-

Figure 6.4
Team Teaching Formats

Team Teaching Approaches

Unit Team   Skill Event   Skill Level

nated the leader (master teacher) and the others are assistant teachers. It is the master teacher's responsibility to organize and delegate duties to the assistants. These duties may include securing specific resources, teaching selected skills, and evaluating students. In the *skill event approach*, teachers select skills in specific units to teach. For example, in a seventh-grade volleyball unit, one physical educator may teach the overhead set, another may teach the underhand set, and a third may teach the serve. In the *skill level approach*, students are grouped by ability and assigned to different teachers accordingly. For example, in a class of 30 eighth-grade students, one physical educator may teach 18 beginner tennis students, while a second may teach 12 advanced students.

Team teaching is an attempt to facilitate learning by reducing the student-to-teacher ratio. Also, the physical education department is able to use the most qualified faculty by allowing them to teach in their area(s) of expertise. As well as providing more effective supervision and management, team teaching exposes students to the talents of more than one teacher. In a unit team format, for example, an assistant teacher can provide more individual help because that instructor does not have responsibility for the entire class. In order for team teaching to be effective, it is essential that physical educators establish interactive relationships that allow them to feel more comfortable teaching alongside their colleagues (Rink, 1993). This approach offers teacher in-service, on-the-job training, an opportunity for professional growth, and the exchange of subject matter knowledge. This is especially advantageous for physical educators early in their careers.

PARAPROFESSIONALS

Paraprofessionals (teacher aides) are personnel hired to assist a teacher. Typically these school employees will have less formal education than a four-year baccalaureate degree. Paraprofessionals normally perform some professional-level functions under the supervision of the teacher, but because of insufficient training and/or experience should not be allowed to assume total class responsibility.

Paraprofessionals can maximize teaching effectiveness by decreasing the student-to-teacher ratio.

Paraprofessionals include both paid individuals and volunteers who share a myriad of titles: teacher aides, teaching assistants, educational assistants, and instructional assistants. Due to increased enrollments and shrinking financial support, schools have dramatically increased their use of paraprofessionals since the 1980s. Paraprofessionals can maximize teaching effectiveness by decreasing the student-to-teacher ratio.

When employing paraprofessionals, school administrators need to consider their (1) qualifications, (2) responsibilities, and (3) training. In physical education, the following minimum qualifications are suggested:

- High-school diploma
- Twenty-one years of age (or older)
- Emotionally mature
- Completed a standard first aid/CPR course
- In good health
- Interested in physical activity

Once the selection criteria are established, the paraprofessional's responsibilities need to be clearly defined and understood by both the physical educator and the aide. Frith (1982) provides a list of guidelines for utilizing paraprofessionals (see table 6.6). The *don'ts* are the responsibility of the certified teacher, whereas the *dos* are suggested tasks for the paraprofessional.

Table 6.6
Paraprofessional Dos and Don'ts

Do	Don't
• Assist with planning (lesson/units)	• Assume the role of a substitute teacher
• Secure necessary equipment	• Conduct unsupervised activity
• Modify activities with teacher	• Make curricular decisions
• Administer selected tests (fitness, skill, etc.)	• Make instructional decisions
• Demonstrate selected skills	• Decide on discipline methods
• Assist in behavior management	• Administer corporal punishment
• Assist in clerical duties (grade exams, record scores, etc.)	• Assign grades
	• Initiate parental contact

Source: Frith, G. H. *The role of the special education paraprofessional: An introductory text.* (1982). Springfield, IL: Charles C. Thomas. Used with permission.

To derive the most from a paraprofessional, some training (professional development) may be necessary. It has been said that paraprofessionals are the fastest growing and least prepared professionals in education. Perhaps the best way to assure that paraprofessionals have the knowledge necessary for supporting a physical education curriculum is pre-service education. With such training, the paraprofessional arrives on the first day of work having completed a program of study covering, at a minimum, (1) the learner, (2) the content, and (3) instructional strategies. Too often, however, paraprofessionals with few or any of these qualifications are employed. If an underqualified individual is hired, the responsibility then falls on the physical education department and/or the school for providing in-service education.

THE SELECTIVE PROGRAM

Traditionally, it had been considered sound practice to require a predetermined program from K–12. This requirement ensured a breadth of experiences for all students. This "we know what's best for you" approach, however, frequently reduced students' motivation to participate. During the formative years from kindergarten through middle school, a required physical education program that will facilitate the student's optimum development and understanding of a variety of games, skills, and movement experiences is recommended. Furthermore, the required program is apt to lead to larger percentages of physically active students. However, the trend today encourages individual choice within the curriculum. Individual choice generates student interest, which leads to continued involvement in physical activity. To achieve this end, a *selective* curriculum is recommended. This program allows students to choose from a listing specific activities in which they have an interest. Students can then concentrate on these selected activities. This approach has the greatest application at the 7–12 grade level. When secondary schools abandon a rigid, structured required physical education program and allow students to choose activities, a better understanding of and greater appreciation for the program emerges.

ABILITY GROUPING

Ability grouping involves placing students in groups according to levels of performance. As early as 1967 the American Medical Association's Committee on Exercise and Physical Fitness promoted the importance of proper grouping in physical education (Chambers, 1988). It was the committee's opinion that some degree of homogeneity needed to be present in order to encourage participation and thereby achieve physical education objectives.

The primary purpose of grouping students with similar skill levels is to create an instructional setting that is most conducive to learning. Students have traditionally been grouped on the basis of chronological age, grade level, interest, and anthropometric assessment. With the increasing range of ability found in coeducational classes, there is a growing need to provide equal opportunity for all students. However, it can be frustrating for the instructor when a wide range of ability exists. Too often, this frustration leads to the practice of engaging almost wholly in recreational games.

Ability grouping, on the other hand, provides the teacher with a better opportunity for meeting the needs of all students. In addition, it allows students to more easily (1) attain success in a movement environment, (2) learn a skill through more effective teaching, and (3) develop leadership skills. By grouping students homogeneously, it is possible to focus instruction at the appropriate level of individual ability. Why should a good tumbler repeat forward rolls just because it is the lesson of the day for everyone in class? When students are ability grouped, high-ability individuals improve more than when they are taught in a mixed-ability class.

Often times, poorly skilled younger children become discouraged as they watch their more skillful classmates. In such a case, ability grouping can give these lesser skilled students an opportunity to succeed in a less competitive atmosphere. One drawback to this is that young children often learn by imitation, and segregating them from their better-skilled peers could deprive them of a significant source of models. Separation also may lead to stereotyping and/or labeling. Because low-ability students do not always benefit from ability grouping, Hellison and Templin (1991) recommend that ability grouping be used only when the physical educator (1) can effectively interact with different groups, (2) is able to validly assess students for placement, and (3) following the initial placement, is flexible about reassigning students as the need arises.

Concerns on how to group and how much time to devote to the grouping of students are legitimate. The use of student self-ratings, teacher rating scales, skill tests, and game performance are all appropriate techniques for judging ability. Ultimately, the decision regarding ability has to be resolved locally, keeping in mind the resources that are available. Certain teachers may employ observation and encouragement to get good results in a mixed group, while other teachers may work more effectively with students who are similar in ability.

PERFORMANCE CONTRACTS

Contracting is a form of individualized study whereby the student embarks on a self-directed learning experience. It allows students to be responsible for their own learning while progressing at their own pace. Contracting is applicable for both psychomotor and cognitive behaviors.

Using performance contracts, the physical educator and the individual student *agree* on a specific behavior in order to fulfill a certain requirement. It is an instructional strategy in which a reward, usually a grade, is awarded to the student at the completion of the contract.

Contracts may be drawn up for a total unit of activity or for a single skill within a unit of activity. In addition, contracts can be used at the elementary school, middle school, and high school levels. Examples of contracts are illustrated in figures 6.5 and 6.6. Students may be allowed to do in-class and/or outside-of-class work on their contracts. Contracting can be a useful strategy, but it is essential that both the teacher and student understand the terms of the contract.

COMPUTER-BASED INSTRUCTION

Educators are constantly searching for the most effective pedagogical method to deliver content, from the long-standing lecture approach to the interactive student participation approach. Perhaps the greatest influence has stemmed from the rapid development and increased utilization of computer-based instruction (CBI). Since its inception in the 1980s, CBI has grown to encompass a variety of instructional components. The two components providing the greatest impact are computer-assisted instruction (CAI) and interactive video instruction (IAV).

CAI utilizes the computer to provide content through an on-screen presentation. This information is typically presented in the form of text and/or graphics. Technological developments have enabled the merging of full-

Figure 6.5
Hoop Contract for Grades 2–3

Name: _____

Grade: _____

Date Completed: _____

Check when you have performed the skill successfully five out of five times.

( ) 1. Spin hoop on one arm.

( ) 2. Spin hoop on the other arm.

( ) 3. Pass hoop from one arm to the other while spinning it.

( ) 4. Hold hoop high over head and drop it so hoop hits the ground without touching the body.

( ) 5. Throw hoop into the air and catch it.

( ) 6. Throw hoop into the air, let it bounce, and catch it.

( ) 7. Throw hoop into the air and catch it on your foot.

( ) 8. Roll hoop on the floor so it will come back to you.

( ) 9. When hoop comes back after a roll, run through it.

( ) 10. When hoop comes back after a roll, jump over it.

Source: Bucher, C. A., & Thaxton, N. A. (1979). *Physical education for children: Movement foundations and experiences.* New York: Macmillan.

Figure 6.6
Basketball Contracts for Grades 7–12

Name: _____

Grade: _____

Date Completed: _____

Note: Contracts 1, 2, 3, 5, and 6 are required.

Contract 1 (10 points). Research and write a paper (two pages) on the history of basketball. The paper must be typed. A minimum of two references must be used, and a bibliography must be included.

Contract 2 (5 points). Explain any <u>3</u> of the following rules and/or fouls:

1. Throw in from out-of-bounds

2. Player positions for free throws

3. Three-second violations

4. Dribbling infractions

5. Charging and blocking

6. Personal and technical fouls

Contract 3 (20 points). Pass a written test with a score of 80 percent or higher. The test will cover basic rules, terminology, and history of basketball.

Contract 4 (10 points). Devise a lead-up game to basketball and include rules, number of players, and the equipment and facilities needed.

(continued)

Contract 5 (10 points). Demonstrate and explain these 7 skills to the teacher:
1. One-hand push pass
2. Two-hand push pass
3. Two-hand bounce
4. Set or push shot
5. Lay-up shot
6. High and low dribble
7. Defensive stance and movement

Contract 6 (35 points). Pass a skills test covering the skills listed in Contract 5. In order to receive the total point value, you must make a score of 90 percent or higher.

Contract 7 (5 points). Play a game of one-on-one basketball with a classmate. By counting each basket as one point, play a 21-point game.

Contract 8 (10 points). Officiate a basketball game during the intramural basketball tournament.

Contract 9 (5 points). Construct a single elimination tournament for eight teams. Explain the tournament, emphasizing why it would be used instead of other types of tournaments.

Contract 10 (5 points). Watch a high school basketball game and a professional basketball game during the next two weeks and write a few paragraphs explaining three rules that are different and three rules that are the same for both games. Make a general statement regarding the basic difference in the games as you view them.

Contract 11 (10 points). Teach a student a basketball skill that he or she cannot perform until that student is able to pass a performance test on that skill.

Contract 12 (5 points). Explain any 2 of the following concepts:
1. "Giving with the ball" as it is caught.
2. Stepping in the direction of the pass.
3. Following through when passing.
4. Putting backspin on the ball when shooting.

Source: Bucher, C. A., & Thaxton, N. A. (1979). *Physical education for children: Movement foundations and experiences.* New York: Macmillan.

motion video, quality audio, and enhanced graphics to produce the more advanced IAV, commonly called the multimedia approach. IAV allows the educator to individualize the content to match the student's developmental needs as well as his or her interests. Computer-based instruction, including CAI and IAV, falls into one of the following three categories:

- Drill and practice
- Tutorial
- Simulation

In drill and practice programs, the content is initially presented and then followed by one or more questions. Following the student's response, the computer will evaluate it and provide immediate feedback. Tutorial programs are similar but call for greater student involvement. Beyond just reading text and answering questions, students can periodically review prior content and/or take an exam. Simulation programs provide real life situations involving student interaction. Simulation programs are most effective in an IAV format because of the ability to show full-motion video.

Although K–12 physical education seldom employs passive lecture methodology, there are some cognitive components of physical education that lend themselves to this approach, including the history, rules, terminology, and strategy of various games and sports. Passive lecture methodology has been presented through both CAI and IAV (Adams, Kandt, Throgmartin, & Waldrop, 1991; Justen, Adams, & Waldrop, 1988; Kerns, 1989). CBI also has been used in the psychomotor domain.

INSTRUCTIONAL STRATEGIES USING STUDENTS

Using students for instructional purposes is a common occurrence in physical education. The transfer of instructional responsibility from the teacher to the student has obvious benefits. Using students allows the physical educator more time for content organization, class supervision, and the utilization of different teaching techniques. These experiences provide an opportunity for students to nurture desirable personality traits like self-confidence, responsibility, and leadership qualities. In addition, student involvement has the potential to:

- Aid the student in developing observation and analysis skills (Rink, 1993)
- Provide a better understanding of the motor skills being taught (Rink, 1993)
- Increase social interaction (Mosston & Ashworth, 1994)
- Increase student learning for all involved (Mosston & Ashworth, 1994)

Two approaches for using students as *teachers* exist—student assistants (SA) and peer teachers (PT). Older students are often used as assistants with younger students; for example, having eighth- and ninth-grade students assist in the primary grades. This approach can be successful in motivating both the younger students being taught and the older students doing the teaching. The implementation of student assistants can enrich the overall physical education program. The responsibilities for SAs include:

- Teaching content selected by the physical educator
- Supervising group activity as assigned by the physical educator
- Providing individual attention
- Assisting in preclass organization (preparing an area for activity)
- Assisting in postclass clean up

Peer teaching involves learners teaching learners, what Mosston and Ashworth (1994) term *reciprocal teaching*. PT can be used for (1) an entire lesson (see table 6.7), (2) part of a lesson (see table 6.8), (3) individualized instruction, or (4) assessment. With individualized instruction the PT explains and/or demonstrates a skill to another student in the same class. It works well because peers can sometimes communicate the material more effectively than the teacher. For best results with individualized instruction, Ernst and Pangrazzi (1996) provide the following guidelines:

Table 6.7
Peer Teaching—Entire Lesson

Gymnastics (Grade 9)

1. The physical educator divided the students into ability groups of four people. Using previously learned floor exercise skills, each student in the group is expected to teach his/her routine to the other members of the group.

2. The physical educator explains that the student who is teaching the routine (peer teacher) is responsible for the quality of performance of the learners and that groups will not be evaluated on the level of difficulty but rather on the following criteria:

   • Clarity of body shape throughout the routine

   • Smoothness of transition from one move to another

   • Control of movement

   • Dynamic quality of execution

3. The peer teacher is encouraged to first demonstrate and explain how each part of the routine is done and then give students practice on each part. When students can do each part with quality, the peer teacher puts the parts together. Groups move at their own pace but are encouraged to practice one routine until it is done successfully before moving onto another routine.

Source: Adapted from Rink, Judith E. (1993). *Teaching physical education for learning.* St. Louis, MO: Mosby. Used with permission.

Table 6.8
Peer Teaching—Individual Skill

Volleyball

1. The physical educator works with the entire class on the volleyball underhand serve. The teacher then divides the class into groups of four people. One student serves the ball from one side of the net, and one student serves the ball from the other side of the net. One student on each side (peer teachers) coaches the server. The coach's job is to check for the following teaching cues that have been given for underhand serve:

   • Using up and back stance with body lean

   • Hitting the ball out of the hand with no toss

   • Finishing with weight on the front foot

2. Each group has a skill card with the cues listed. Coaches are told to look for only one cue each time the ball is served and then tell the server whether that cue has been observed.

Source: Adapted from Rink, Judith E. (1993). *Teaching physical education for learning.* St. Louis, MO: Mosby. Used with permission.

- When psychomotor outcomes are desired, low-skilled learners should be paired with a high-skilled learner.
- When affective outcomes are desired, the learners should be allowed to select their peer.
- When cognitive outcomes are desired, the learners should be encouraged to select a compatible peer.
- When competition is the desired outcome, learners should be paired by size, strength, and ability.

Peer assessment occurs when one student assesses the performance of a classmate (Melograno, 1997). This is done formatively. Johnson (2004) states that peer assessment enhances the learning experience and affects achievement. Its best application requires a classmate to assess a peer using the same skill test that will be used for the summative assessment (see chapter 11 for a discussion of formative and summative evaluations). Johnson (2004) adds that when used this way, peer assessment may:

- Provide additional practice for the performer
- Provide individual feedback from the peer
- Provide exposure to the summative test

SUMMARY

1. Curriculum reform is best accomplished when all individuals—including administrators, teachers, parents, and students—become involved in the process.
2. A curriculum guide presents the physical education program in written form and provides a framework from which to review and/or revise.
3. The development of a curriculum guide is the responsibility of all individuals involved, including administration, teachers, parents, and students.
4. Before the curriculum guide is written it is recommended that example guides be obtained from public schools and/or state departments of education.
5. Scope refers to *what* is to be taught and sequence refers to *when* it is to be taught.
6. To guide the development of scope and sequence, Bookwalter provides five criteria—progression, variety, seasonality, feasibility, and practice for mastery.
7. When scheduling the physical education curriculum, two organizational patterns are typically used—traditional and flexible.
8. The cycle plan allows for more intensive study of a specific activity.

9. Both indoor and outdoor facilities are used for the instruction of physical education.

10. Team teaching is an approach to staffing that can be beneficial to students by reducing the student-to-teacher ratio, providing more effective supervision and management, and exposing students to the talents of more than one teacher.

11. The use of paraprofessionals has grown in these times of shrinking resources. Of specific concern is the selection, training, and role delineation of paraprofessionals.

12. Allowing secondary school students to select what activities they want to be taught may lead to a better understanding of and appreciation for physical education.

13. Homogeneity in physical education classes can be accomplished through ability grouping.

14. Two approaches to individualized study in a physical education curriculum are contracting and computer-based instruction.

15. Students can be useful for instruction either as teaching assistants or peer teachers.

QUESTIONS AND LEARNING ACTIVITIES

1. Interview the physical education director of a large school. Ask how the total curriculum was developed and how the physical education faculty carries it out. Examine the program for seasonal content, time allotment, variety of activities, and electives available for both boys and girls.

2. Examine literature pertaining to ability grouping, computer-based instruction, performance contracts, and team teaching. List two or three advantages and disadvantages for each. Be prepared to discuss what implications need to be considered when using one of these innovations.

3. Provide descriptions of scope and sequence. Visit selected schools to secure a copy of their curriculum guides, and identify the scope and sequence of their activities.

4. Suppose you are employed by a large urban school system as the director of physical education. You have been asked to lead a task force to review the current K–9 curriculum. Indicate what steps you would take to accomplish this task.

5. Using the following criteria for the organization of physical education activities, determine which criterion is exemplified in the following statements. Place the letter(s) of your response in the space provided.

A. Progression   D. Feasibility

B. Variety   E. Practice for Mastery

C. Seasonality

_____ 1. Adapting the choice of activity to suit weather and the prevailing interest in sports.

_____ 2. Allowing more than three volleys in volleyball for beginners.

_____ 3. Assigning a series of forward rolls after the single roll is learned.

_____ 4. Avoiding the boring repetition of activities.

_____ 5. The necessary equipment is available for the activity.

_____ 6. Extensive rather than intensive treatment.

_____ 7. Gradual increase in performance, difficulty, or complexity in accordance with the individual's readiness.

_____ 8. Intensive rather than extensive treatment.

_____ 9. Lowering basketball goals for the intermediate grades.

_____ 10. Selecting activities that contribute to all of the objectives.

_____ 11. Sufficient drills to assure learning.

_____ 12. Planning a specific activity for early spring.

_____ 13. Repeating the same stunt or exercise reasonably often.

_____ 14. Selecting activities that are practical under the conditions.

_____ 15. Staying with an activity until ease of function is acquired.

REFERENCES

Adams, T., Kandt, G., Throgmartin, D., & Waldrop, P. (1991). Computer-assisted instruction vs. lecture methods in teaching the rules of golf. *The Physical Educator, 48*(3), 146–150.

Annarino, A., Cowell, C., & Hazelton, H. (1986). *Curriculum theory and design in physical education.* Long Grove, IL: Waveland Press.

Bookwalter, K. (1964). *Physical education in secondary schools.* New York: The Center for Applied Resources in Education.

Bukowski, B., & Stinson, A. (2000). Physical educators' perceptions of block scheduling in secondary physical education. *Journal of Physical Education, Recreation, and Dance, 71*(1), 53–57.

Chambers, R. (1988). Legal and practical issues for grouping students in physical education classes. *The Physical Educator, 45*(4), 180–185.

Ernst, M., & Pangrazzi, B. (1996). Two by two: The benefits of peer teaching. *Teaching Secondary Physical Education* (December), 21–22, 25.

Frith, G. (1982). *The role of the special education paraprofessional: An introductory text.* Springfield, IL: Charles C. Thomas.

Greenwood, M., & Stillwell, J. (1999). State agency curriculum material for physical education. *The Physical Educator, 56*(3), 155–158.

Hellison, D., & Templin, T. (1991). *A reflective approach to teaching physical education*. Champaign, IL: Human Kinetics.

Jewett, A., Bain, L., & Ennis, C. (1995). *The curriculum process in physical education*. Dubuque, IA: Brown & Benchmark.

Johnson, R. (2004). Peer assessments in physical education. *Journal of Physical Education, Recreation, and Dance, 75*(8), 33–40.

Justen, J., Adams, T., & Waldrop, P. (1988). Effects of small group versus individual user computer-assisted instruction on student achievement. *Educational Technology, 28*(2), 50–52.

Kelly, L. (1987). Computer-assisted instruction: Applications for physical education. *Journal of Physical Education, Recreation, and Dance, 58*(4), 74–79.

Kerns, M. (1989). The effectiveness of computer-assisted instruction in teaching tennis rules and strategies. *Journal of Teaching in Physical Education, 8,* 170–176.

Melograno, V. (1996). *Designing the physical education curriculum*. Champaign, IL: Human Kinetics.

Melograno, V. (1997). Integrating assessment into physical education. *Journal of Physical Education, Recreation, and Dance, 68*(7), 34–37.

Mosston, M., & Ashworth, S. (1994). *Teaching physical education*. New York: Macmillan.

Rettig, M., & Canady, R. (1999). The effects of block scheduling. *The School Administrator, 56*(3), 14–16, 18–20.

Rink, J. (1993). *Teaching physical education for learning*. St. Louis, MO: Mosby.

7

THE ELEMENTARY PHYSICAL EDUCATION PROGRAM, K–6

Outcomes

After reading and studying this chapter, you should be able to:
- Define

Basic rhythms	*Nonlocomotor skills*
Body awareness	*Relationships*
Cooperative games	*Relays*
Creative rhythms	*Self-testing*
Folk dance	*Singing rhythms*
Lead-up games	*Social dance*
Locomotor skills	*Spatial awareness*
Low-organized games	*Sports*
Manipulative skills	*Square dance*
Movement education	*Traditional and contemporary dance*
Movement qualities	

- Identify the six content categories for an elementary physical education program.
- Describe the physical/physiological characteristics of elementary school students.
- Describe the intellectual characteristics of elementary school students.
- Describe the social/emotional characteristics of elementary school students.
- Provide specific examples in each of the six elementary physical education content categories.
- Discuss the place and importance of games in an elementary physical education curriculum.
- Discuss the time allotment for each of the elementary physical education content areas for grades K–6.
- Select physical education activities for a K–6 program based on the students' physical, intellectual, and social/emotional characteristics.

143

The concept that children's play is as old as culture itself is supported by the disciplines of the classics, archaeological findings, and the relics or monuments of ancient civilization. From Rousseau to Piaget it has been called priceless—the children's way of life. Play is what children do when they are not eating, sleeping, or complying with the wishes of adults (Gallahue & Donnelly, 2003). It is through play that children learn about themselves and their movement capabilities. More importantly, children learn about the people and the world around them by means of playing. Children have not waited for others to teach them how to play. Rather, children have taught themselves in a manner that is serious and meaningful. This "self-teaching" is a profound dimension because the naturalness, innocence, and lack of sophistication that exist in children allow us to observe how children feel not only about an activity but also about each other while at play.

CHILDREN MOVING

One goal of children's play, much of which is in a movement setting, is the efficiency of movement. The movement-learning process is partly an intellectual one because developing both perceptual-motor skills and rich sensory experiences is paramount to young people's understanding of themselves and their adjustment to the world of people, things, and ideas. It is through the early acts of moving, both individually and collectively, that children learn to associate competence with social acceptance. This association is an important dimension in a child's development. Research supporting this concept shows that there is a positive relationship between motor-skill ability and social status, self-concept, and popularity.

Movement, more specifically how a person moves, is the essence of existence itself to a child. The self-discovery, freedom, kinesthetic feeling, and sheer enjoyment of movement signify much to a child. Thus, educators should strive to provide a balanced physical education curriculum at the elementary school level. This curriculum should provide numerous ways—individually, with a partner, and in groups—for children to learn the fundamental locomotor, nonlocomotor, and manipulative skills. These skills are shown in table 7.1.

Table 7.1
Fundamental Skills

Locomotor	Nonlocomotor	Manipulative
Gallop	Bend/Stretch	Catch
Hop	Push/Pull	Dribble
Jump	Rise/Fall	Kick
Leap	Swing/Sway	Strike
Run	Twist/Turn	Throw
Skip		
Slide		
Walk		

THE ELEMENTARY SCHOOL CHILD

Both program planning and effective teaching require a familiarity with the signs of growth and development in the psychomotor, cognitive, and affective domains. From this perspective it is better to group children rather than attempt to pinpoint behavior at one particular year. In addition, a time never exists when all boys and girls in a particular class are at the same growth age.

Developmental characteristics are presented to provide a more complete view of the elementary school child (see tables 7.2, 7.3, and 7.4). A thorough understanding of these characteristics and their implications will make it easier and more educationally sound when selecting the elementary physical education content. It is important to remember who will be taught.

Table 7.2
Characteristics of Students Ages 5–7

Physical/Physiological

- Growth is relatively slow but steady in both height and weight as students grow from two to three inches and gain from five to seven pounds per year, respectively.
- The resting heart rate may exceed ninety beats per minute, while the respiration rate is eighteen breaths per minute.
- Endurance is poor, causing students to fatigue easily.
- The development of both strength and muscular endurance is steady.
- Reaction time is slow, showing gradual improvement with an increase in age.
- Perceptual-motor abilities are incomplete but developing. Because the eyes are slow to focus and students are farsighted at this age, hand-eye coordination is poor.
- Gross motor control is developing rapidly and at a more advanced rate than fine motor control.
- The development of fundamental movement skills, both locomotor and manipulative, is occurring rapidly.
- Rhythmic skills show a gradual increase in quality.
- Bones, still developing, are relatively soft and pliable.
- Pelvic tilt is evident in the early stages of this developmental group.
- Static (stationary) and dynamic (moving) balance show steady improvement.
- Students are active and energetic.
- Gender differences are insignificant.

Intellectual/Social/Emotional

- Attention span is short.
- Play is more individual/partner rather than group oriented.
- Students are creative, imaginative, and imitative.
- Students are curious and adventurous.
- Students are shy and self-conscious.

(continued)

- Students are egocentric, while most behaviors are self-satisfying.
- Adult approval becomes important as they develop a desire to please.
- Students do not accept criticism well.
- Self-concept is rapidly developing.
- The ability to reason, exercise judgment, and solve problems is developing, but intuition is used more than logic in the thinking process.

Implications

- Provide daily, vigorous activity while limiting extended periods of inactivity.
- Provide a variety of large-muscle movement experiences.
- Provide activities of short duration.
- Make certain that initial instruction includes shorter distances and lower speeds.
- Provide frequent rest intervals, if necessary.
- Provide activities involving a change in speed, force, and level of movement.
- Provide activities requiring balance and agility, employing both large and small apparatus.
- Incorporate a variety of rhythmic activities, including creative dance experiences.
- Provide a variety of manipulative activities. Use large (seven-inch) balls for ease in tracking.
- Provide wholesome coeducational activities.
- Provide abdominal strengthening exercises and activities.
- Provide activities employing imagery, drama, and problem solving to foster creativity.
- Incorporate bilateral skills (e.g., skipping and galloping) once unilateral skills have been some-what well developed.
- Stress the development of fundamental locomotor and manipulative skills that progress from simple to complex.
- Emphasize process or the quality of the movement, rather than product or the end result of the movement. Correct execution should be the primary goal.
- Provide for individual differences by letting students progress at their own rates.
- Take time to explain selected components of the class.
- Include ample, genuine positive reinforcement.
- Limit negative comments.
- Continually stress safety.

Sources: Gallahue & Donnelly, 2003; Kirchner & Fishburne, 1995; Wall & Murray, 1994.

Table 7.3
Characteristics of Students Ages 8–9

Physical/Physiological

- Height and weight gains continue to be steady.
- The heart and lungs continue to develop slowly with a steady decrease in both resting rates.
- Strength improvement is steady as muscle mass will nearly double from ages five to ten.
- Reaction time is slow but becomes established by the end of this stage.
- Hand-eye coordination, balance, and other perceptual-motor abilities are improving.

(continued)

- Fundamental movements are well developed and nearing refinement.
- Poor posture is common, more often in girls.
- Rhythmic skills are nearing refinement.
- Hand preference is firmly established.
- Flexibility has decreased slightly, especially in boys.
- Students are still active and easily fatigued.

Intellectual/Social/Emotional

- Attention span continues to increase. At the end of this stage, children often become focused.
- Students seek independence.
- A shift from individual/partner to group activities occurs.
- Students remain curious, adventurous, and eager to know why.
- Students have become both self-conscious and self-critical.
- A strong desire to please adults exists. As a result, children are responsive to authority.
- Students exhibit an eagerness to learn.
- Students are eager to try new activities but become frustrated if the required movements are beyond their capabilities.
- Group identity, loyalty, and acceptance become important.
- Students exhibit fear of failure, especially in selected movement skills.
- Students are not capable of abstract thinking.
- Students become aggressive, boastful, and sometimes argumentative.

Implications

- Continue to provide daily, vigorous activity.
- Provide exercises and/or activities for improving flexibility.
- Provide more complex balance and agility activities.
- Provide more complex locomotor and basic game skill activities.
- Introduce a variety of sport-specific skills through drills and lead-up games.
- Provide activities that incorporate music to enhance the understanding of rhythmic components and to refine coordination.
- Provide more complex rhythmic skills including structured folk, square, and contemporary dance.
- Continue to provide activities that stress the antigravity muscles.
- Encourage students to maintain proper posture. Also instruct them on proper body mechanics.
- Provide group activities to enhance both interaction and appropriate social behavior.
- Provide abundant opportunities for encouragement and positive reinforcement. Assure children they are important and valued.
- Provide a variety of self-testing activities so students can independently determine their skill level.
- Focus on skill acquisition while allowing ample time for practice.
- Integrate academic concepts with movement.
- Encourage children to think before they become involved in an activity.
- Provide concrete examples and situations for enhancing cognition.

Sources: Gallahue & Donnelly, 2003; Kirchner & Fishburne, 1995; Wall & Murray, 1994.

Table 7.4
Characteristics of Students Ages 10–12

Physical/Physiological

- At the onset of puberty and into adolescence, children experience a rapid growth spurt. Girls will reach puberty between ages 10 to 11 while boys will reach puberty around ages 12 to 13.
- Heart and lungs have increased in size, which is proportionate to an increase in overall growth. As a result, cardiovascular endurance has increased.
- Reaction time is improved.
- Overall coordination, although continually developing, may slowly regress due to a rapid growth spurt.
- Hand-eye coordination is fully developed.
- Most of the fundamental movements are refined and, as a result, have become overlearned.
- Gender differences appear with girls as they become taller and more physiologically advanced.
- Girls may be more developed in selected sport-specific skills.

Intellectual/Social/Emotional

- Students continue to seek independence.
- Interest in specific movement activities broadens.
- A concern for fitness and refinement of sport-specific skills is keen.
- Students are self-conscious of motor inadequacies.
- Group loyalty is strong as indicated by the formation of cliques.
- Students seek social acceptance from their peers rather than their teachers.
- Competition becomes important.
- Difficulty in controlling emotions occurs, often causing students to overreact.
- Sexual modesty and an interest in the opposite sex are developing, as is sexual antagonism (e.g., hitting, teasing, and chasing).
- An interest in appearance is apparent.
- Students enjoy being challenged, both physically and mentally.
- The activity interests of boys and girls begin to diverge.
- Fads in food, clothes, and movement experiences often influence interests.
- Students exhibit a capacity for self-evaluation.
- Individual differences are many and varied.
- Students have developed an ability to deal with abstract concepts.

Implications

- Continue to provide daily, vigorous activity.
- Provide strenuous activities to work the muscles, heart, and lungs.
- Continue to provide a variety of perceptual-motor activities.
- Provide ample opportunities for students to refine all of the basic game skills. In addition, increase opportunities for students to combine selected fundamental skills so their movements become efficient.
- Promote self-reliance by exposing students to experiences requiring responsibility.
- Allow leadership opportunities.

(continued)

- Provide activities that foster fair play and sportsmanship.
- Continue to provide group activities to enhance both interaction and appropriate social behavior.
- Provide challenging experiences.
- Continue to integrate academic concepts with movement.
- Provide experiences that are developmentally appropriate and foster skill acquisition.
- Stress accuracy and form during skill instruction.
- Incorporate timely activities that have contemporary interest.
- Provide activities that are geared to both the needs and interests of the students.
- Provide assistance so students can more easily make the transition from the fundamental movement phase to the sport-specific movement phase.
- Encourage students to participate in youth sport activities that are developmentally appropriate.

Sources: Gallahue & Donnelly, 2003; Kirchner & Fishburne, 1995; Wall & Murray, 1994.

CURRICULUM CONTENT FOR THE ELEMENTARY GRADES

If the objectives of the curriculum (see chapter 2) are to be met, a broad scope of physical education offerings must be available. Figure 7.1 shows the scope of the physical education curriculum for a K–6 elementary school. Each of the content areas shown in this figure contains movement experiences that can help children attain their total development. In addition, content areas shown can easily be adapted to fit most schools with the

Figure 7.1
Elementary Physical Education Content

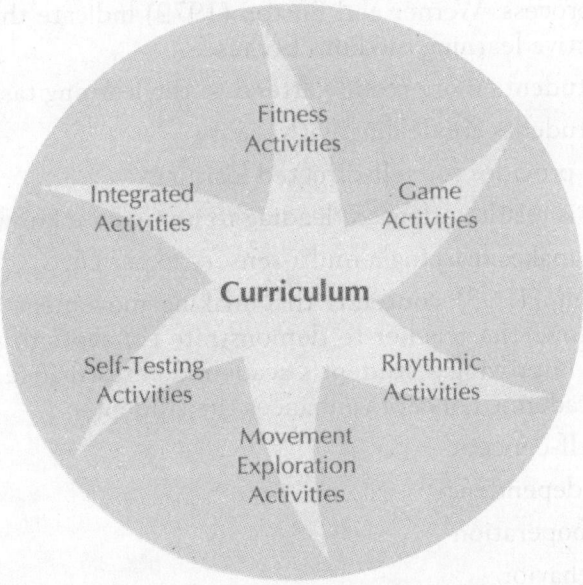

K–6 organizational structure. A discussion of each category is provided for a better understanding of the elementary physical education scope.

Fitness Activities

Even though many of the activities in the physical education curriculum contribute in some way to the development and maintenance of health-related fitness, it is still necessary to include fitness as a separate content area. Moreover, since the formation of the President's Council on Physical Fitness under the Eisenhower administration in 1956, a vigorous, cooperative effort has taken place in this country to raise the physical capacity of our youth. As long as both men and women need the physical capacity to perform a day's work and have a reserve to recreate and/or meet any emergency that may arise, our schools need to be seriously concerned with the topic of health-related fitness.

A person's level of fitness can be enhanced through a variety of content, like games, rhythmic activities, and gymnastics. However, fitness activities are designed to develop and maintain a student's aerobic capacity, strength, muscular endurance, flexibility, and a favorable body composition. This category includes such specific activities as jogging, weight training, jumping rope, and free exercise. For a more detailed description of free exercises, see *More Fitness Exercises for Children* by Stillwell and Stockard (1988).

Integrated Activities

Integrated activities are those that use movement as a medium for academic learning. This active approach to academic learning has been termed "academotion" by the authors. It is based on the belief that learning is an active process. Werner and Burton (1979) indicate that physical activity is an effective learning medium because:

- Students more readily attend to the learning task
- Students are dealing with reality
- It provides for self-directed learning
- It is results-oriented, leading to immediate knowledge of results
- It makes learning a multi-sensory experience

Grant (1995) contends that making movement a part of all subject areas allows the teacher to demonstrate concepts to students, concretely. Beyond improving a student's academic performance, using movement to teach academic concepts enhances the student's:

- Self-concept
- Independence
- Cooperation
- Behavior

Five integrated activities for use with elementary students are described in the following sections.

I'm a letter. With students in a random/scattered formation, ask them to make letters of the alphabet using their bodies. Ask them to make:

- The first letter of their first and last names
- The last letter of the first and last names
- Any vowel
- The first letter of the capital of Alabama, California, etc.
- Any letter rhyming with "C"

Repeat the activities with the children in pairs, requiring the partners to make one letter together. (Note: These activities can be done with the students standing or lying.)

I'm a number. With students in a random/scattered formation, ask them to make single-digit numbers using their bodies. Ask them to make:

- The numbers 1, 2, etc.
- Their favorite number
- The number of brothers and/or sisters they have
- The sum of $1 + 2, 2 + 3$, etc.
- The difference of $5 - 4, 5 - 2, 3 - 1$, etc.

Repeat the activities with the children in pairs, requiring the partners to make the single digit number together. (Note: These activities can be done with the students standing or lying.)

Let us spell it. With students in groups of three, ask them to spell selected three-letter words, with each making a separate letter.

Catch and spell. With students in pairs sitting 10–15 feet apart, have them spell words while playing catch. As each toss is made, a letter of the word is pronounced out loud by each.

Movement in a word or two. With students in a random/scattered formation, show them one-word flash cards containing a verb. Ask them to move accordingly. Suggested verbs include:

- Run
- Skip
- Gallop
- Swim
- Fly

A second flash card containing an adverb can be added. Suggested adverbs require the students to:

- Run *slowly*
- Skip *happily*
- Gallop *quickly*
- Swim *cautiously*
- Fly *silently*

Game Activities

Games have a significant role in elementary physical education programs not only because of their popularity with children but also because of their potential developmental value. When used properly, games become an important educational tool. Because of their possible contributions to physical, social, and recreational objectives, games must be chosen carefully. In particular, games must be taught properly and built into the curriculum at an appropriate grade level. A carefully selected program of games can aid in the development of (1) fundamental movement skills such as running, jumping, and skipping; (2) basic games skills like throwing, catching, and kicking; (3) selected components of fitness; and (4) acceptable social behavior.

Games may be classified in a variety of ways. Perhaps the most common classification includes low-organized games, relays, cooperative games, creative games, lead-up games, and sports.

Low-organized games. Low-organization games have few rules; require little, if any, equipment; may be played following a brief explanation; and may be easily modified. Five examples of low-organized games are presented in figures 7.2, 7.3, 7.4, 7.5, and 7.6.

Relays. Relays are small-group, team-oriented activities. They provide an organizational format that is different from other games in that relays aim to develop fundamental movement skills. Five example relays are presented in figures 7.7, 7.8, 7.9, 7.10, and 7.11.

Cooperative games. Cooperative games have become popular primarily through the work of Orlick (1977, 1978, and 1982). Popularity has grown in recent years because cooperative games are nonthreatening and noncompetitive and not only emphasize cooperation but also require it for successful participation. Orlick (1982) describes the benefits of this cooperative approach to game playing by stating that it provides an opportunity for children to play *with* each other, rather than *against* one another. Children play to overcome a challenge, not an opponent. These games allow the participants, merely by their structure, to enjoy the play experience itself. More specifically, cooperative games provide freedom from competition and exclusion and freedom to choose and create. Cooperative games are classified as being (1) games without losers, (2) collective score games, or (3) reversal games.

Figure 7.2
Crows and Cranes

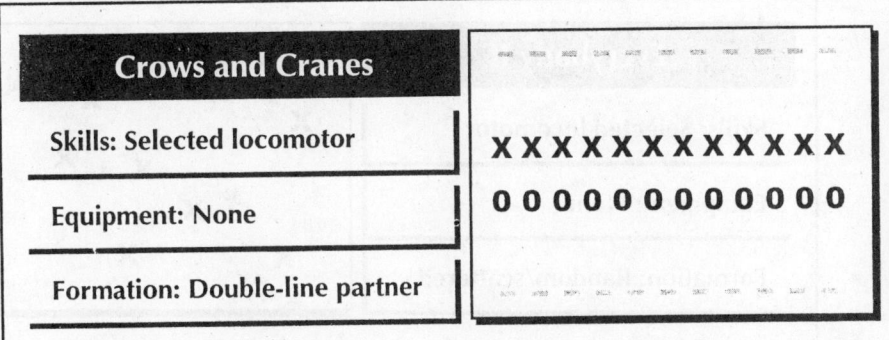

Crows and Cranes

Skills: Selected locomotor

Equipment: None

Formation: Double-line partner

Students are divided into two equal groups facing each other as shown. One group represents the crows and the other group represents the cranes. When the teacher says "crows," the crows turn and run toward their free line. The cranes chase the crows trying to tag them before they cross the free line. If tagged, the student joins the other group. This becomes a dramatic game as students listen carefully for the teacher's cue of "Crrr . . . ows" or "Crrr . . . anes."

Figure 7.3
Partner Tag

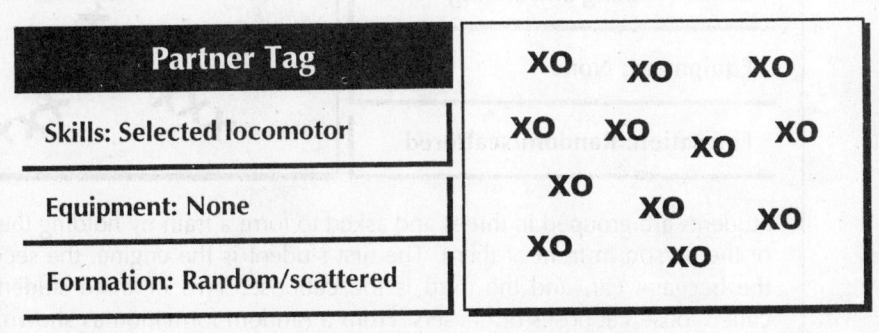

Partner Tag

Skills: Selected locomotor

Equipment: None

Formation: Random/scattered

Students are paired in a random formation, as shown. In each pair, one student is "it." On the command, the students who are it must count to five while their partner is free to run. If tagged, the partner becomes it and must count to five as the other person flees.

Figure 7.4
Everybody It

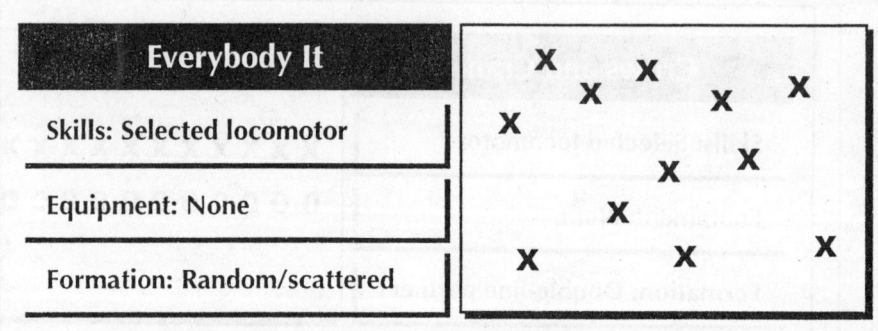

Everybody It

Skills: Selected locomotor

Equipment: None

Formation: Random/scattered

Begin with a random formation as shown. Everyone is "it." On the teacher's command, each student (1) attempts to tag as many other students as possible, and (2) moves to prevent from being tagged by other students.

Figure 7.5
Loose Caboose

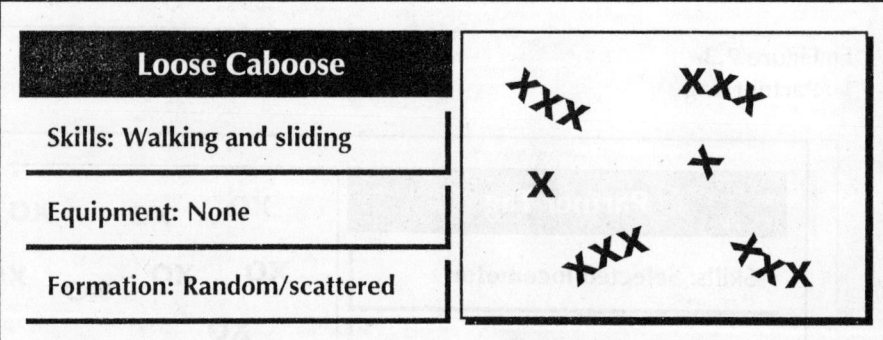

Loose Caboose

Skills: Walking and sliding

Equipment: None

Formation: Random/scattered

Students are grouped in threes and asked to form a train by holding the waist of the person in front of them. The first student is the engine, the second is the baggage car, and the third is the caboose. Two or three students are called loose cabooses or chasers. From a random formation as shown, each train moves to prevent its caboose from being tagged. However, if tagged, the engine must break away to become a new loose caboose.

Figure 7.6
Busy Bee

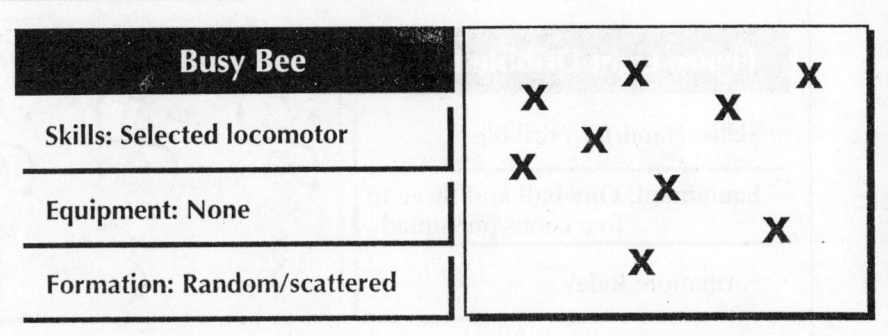

Busy Bee

Skills: Selected locomotor

Equipment: None

Formation: Random/scattered

From a random formation as shown, students buzz as they walk freely in the play area. The teacher calls out a specific command, such as back-to-back. Students move to find a partner and stand back-to-back. The teacher then calls out "Busy Bee." On the command, the students move from their partner and buzz as they again walk about the play area. The game continues as the teacher alternates partner commands, such as hip-to-hip, foot-to-foot, and elbow-to-elbow, with the Busy Bee commands.

Figure 7.7
Locomotor Relay

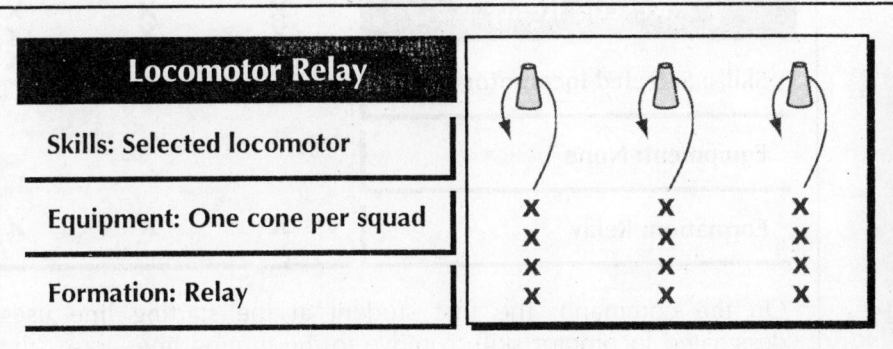

Locomotor Relay

Skills: Selected locomotor

Equipment: One cone per squad

Formation: Relay

On the command, the first student in each squad uses the designated locomotor skill (i.e., walking, skipping, running) to advance toward the cone. The student then goes around the cone, and comes back to the starting line. The student touches the next person in line who repeats this process. This procedure continues until each member of the squad has had a turn.

Figure 7.8
Figure Eight Dribble Relay

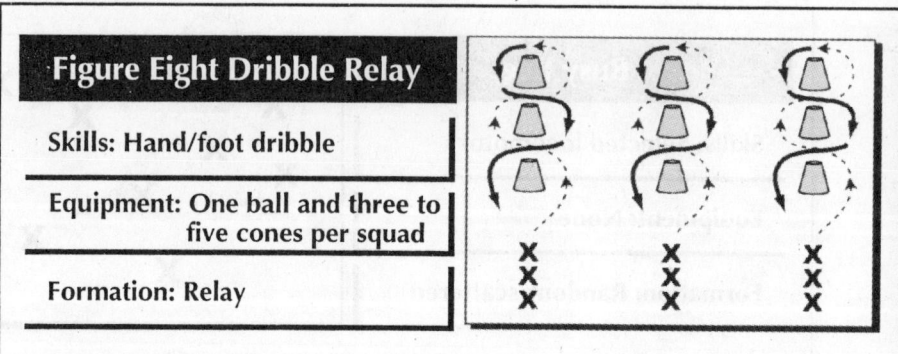

Figure Eight Dribble Relay

Skills: Hand/foot dribble

Equipment: One ball and three to five cones per squad

Formation: Relay

On the command, the first student in each squad weaves around the cones in a figure eight pattern while dribbling a ball with the hand or foot. On returning to the starting line, he/she gives the ball to the next person in line who repeats this process. This procedure continues until each member of the squad has had a turn.

Figure 7.9
Rescue Relay

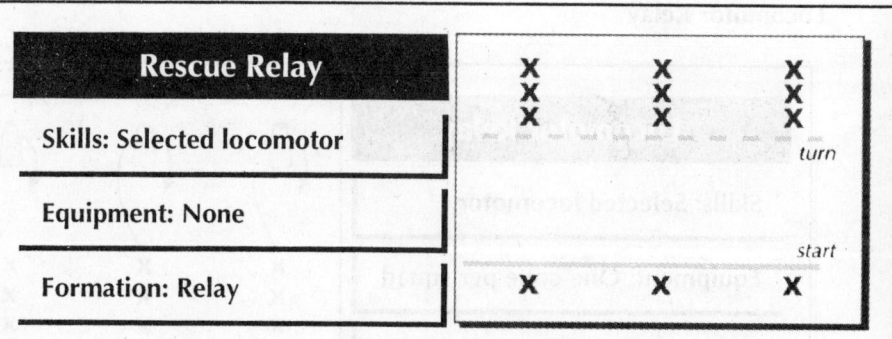

Rescue Relay

Skills: Selected locomotor

Equipment: None

Formation: Relay

On the command, the first student at the starting line uses the designated locomotor skill to move to the turning line, grasps the first squad member's hand, and returns to the starting line (rescuing the squad member). The rescued student then takes a turn and uses the designated locomotor skill to travel back to the turning line to rescue the next person in line. This procedure continues until the last student is rescued and the entire squad is assembled behind the starting line.

Figure 7.10
Shuttle Pass Relay

Shuttle Pass Relay	
Skills: Passing and catching	
Equipment: One ball per squad	
Formation: Shuttle	

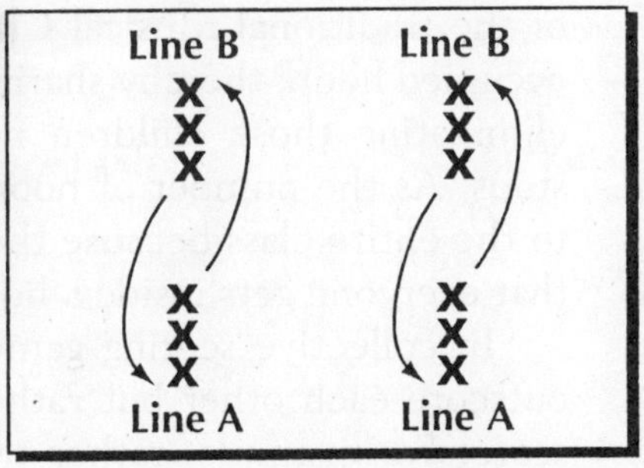

On the command, the first student in line A passes (bounce, chest, or overhead) the ball to the first person in line B. After releasing the ball, the student moves to the end of the approaching line. After catching the ball, the first student in line B passes the ball to the next person in line A. This procedure continues for a designated amount of time or for a designated number of passes.

Figure 7.11
Clean Up Your Plate Relay

Clean Up Your Plate Relay	
Skills: Selected locomotor	
Equipment: One beanbag per student and one hoop per squad	
Formation: Relay	

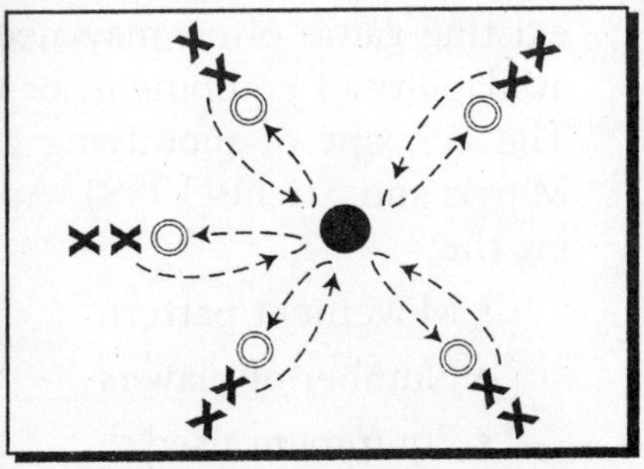

Students are divided into 5 squads and placed as shown. All of the beanbags are placed in the center circle. On the command, the first student in each squad uses the designated locomotor skill to move to the center circle (the plate). The student then picks up one beanbag, returns to the starting position, places the beanbag in his/her hoop, and tags the next student in line who repeats this process. This procedure continues until each student has had a turn and the plate is clean.

Games without losers require students to work cooperatively toward accomplishing a specific task. These games are an alternative to the various elimination games used in elementary physical education. The game Musical Hoops (see figure 7.12) is an example. Musical Hoops is a modification of the traditional Musical Chairs in that children need to step inside any occupied hoop, thereby sharing it with one or more classmates, rather than eliminating those children not standing inside a hoop when the music stops. As the number of hoops diminishes, the game presents a challenge to the entire class because the children must work cooperatively to assure that everyone gets inside a hoop.

In collective scoring games the individuals/teams do not compete to outscore each other but rather work cooperatively to attain a designated score. For example, rather than compete individually in Beanbag Horseshoes (see figure 7.13) to outscore an opponent, the points for each student are totaled to either beat a previous score or reach a predetermined score. In a team game such as Newcomb Serve (see figure 7.14), the students collectively try to score 15 consecutive catches.

In reversal games, children are rotated from squad to squad during a game so that emphasis is placed on each child's performance rather than the team's outcome. For example, the game of Four Goal Soccer (see figure 7.15) is modified by having the scoring player change teams every time he/she scores. Therefore, if a player on team A scores, he/she immediately becomes a member of team B. This presents an immediate challenge to team A, as they have one less team member.

Creative games.   Creative games are derived from one of two sources—by modifying an existing game or developing an original game. An existing game often may need to be modified because of class size, the availability of equipment, or to better meet the objective(s) of the lesson. The concept of modifying games was presented by Morris (1980) and Morris and Stiehl (1989). Appropriate game modifications include changing the:

- Movement pattern
- Number of players
- Equipment used
- Formation
- Space
- Time allotment
- Scoring procedures

Gabbard, LeBlanc, and Lowy (1994) describe original games as those created by the teacher, the teacher and the students, or by the students alone. In order for the teacher-student and the student-alone approaches to be effective, everyone involved needs an understanding of what should

Figure 7.12
Musical Hoops

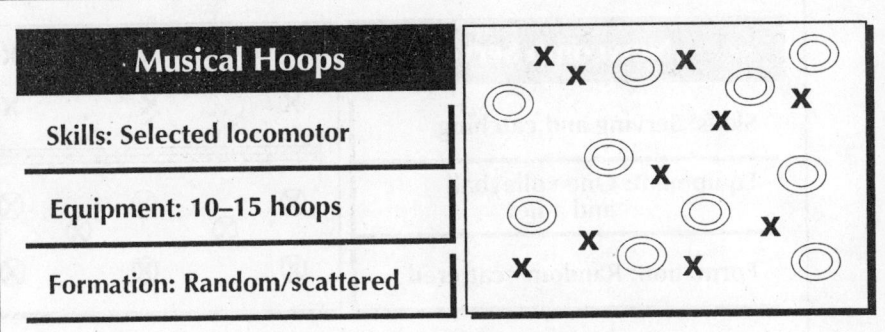

· Musical Hoops

Skills: Selected locomotor

Equipment: 10–15 hoops

Formation: Random/scattered

Ten to fifteen hoops are dispersed throughout the play area. The children, in a random/scattered formation, walk, skip, etc., throughout the play area until the music stops. At that point the children all try to fit themselves inside any of the hoops. The game is continued as one to two hoops are removed each time the music stops.

Figure 7.13
Beanbag Horseshoes

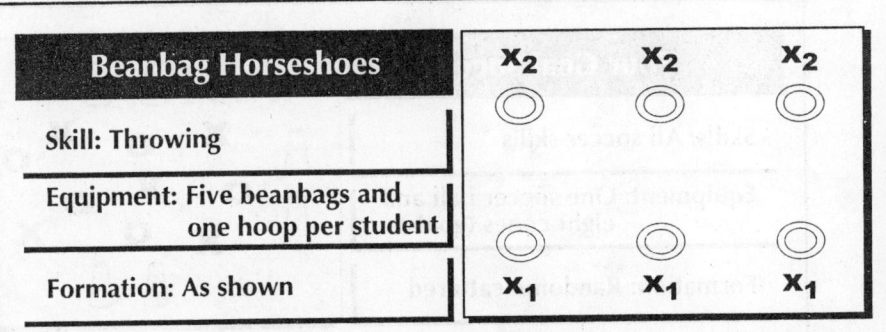

Beanbag Horseshoes

Skill: Throwing

Equipment: Five beanbags and one hoop per student

Formation: As shown

Player X_1, standing behind a hoop, uses an underhand toss to try to land each one of five beanbags inside player X_2's hoop. When player X_1 completes the fifth attempt, player X_2 takes a turn. One point is awarded for each beanbag that lands inside the hoop. The score is the total between partners.

Figure 7.14
Newcomb Serve

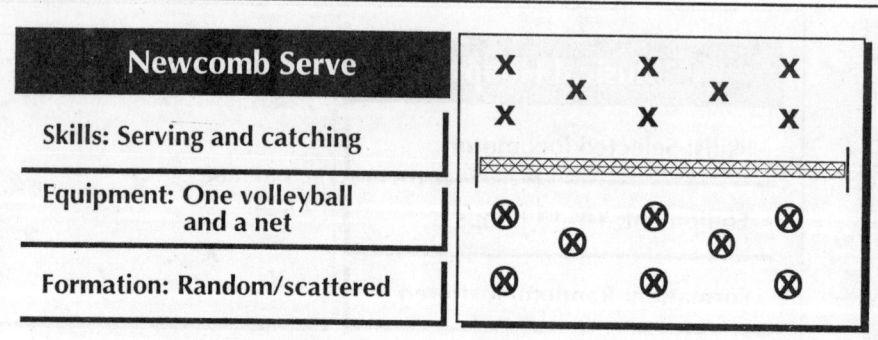

Two teams of six to eight players are positioned as shown. The ball is served from any position. Players on the other side must catch the ball in the air. If caught, it must be served back over the net. Play continues until the ball touches the floor. If the ball is missed, the score reverts to zero. The game is played to fifteen points (fifteen consecutive catches).

Figure 7.15
Four Goal Soccer

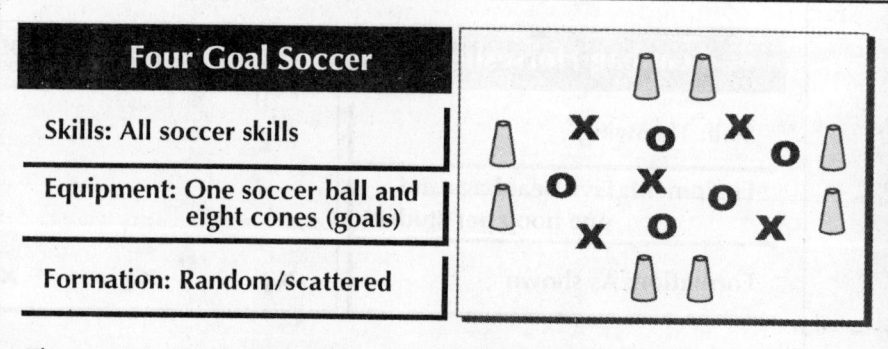

The court is set up with eight cones that form four goals as shown. Two teams of five to eight players each are spread over the playing area. There are no goal keepers as each team attacks and defends two goals. Any player who scores a goal immediately becomes a member of the other team. Soccer rules apply with no officials or corner kicks.

be learned from the game. When working cooperatively on a creative game, the teacher should act as a facilitator by presenting general guidelines for the game. The teacher and students then work together to determine the game specifics, including the rules, equipment, boundaries, and so on. For student-designed games, the teacher's primary role is to supervise and maintain a safe environment. Graham (1977) provides the following guidelines when using student-designed games:

- Progress gradually. More structure is needed at the outset but as students become more adept, the teacher should lessen the imposed structure.

- Limit interference. Allow students to make meaningful decisions in order to promote responsibility.

- Allow students to enforce their own game rules. If allowed the opportunity to make the rules, students should then be involved with enforcing them.

- Be patient. The creative process does not unfold quickly. Remember that the process is just as important as the product. Once students have mastered the general idea of the game components and how to manipulate them, the quality of responses will increase.

- Be flexible while encouraging students to be creative. The primary rule is that the game can always be changed.

Lead-up games.   Lead-up or modified games deserve special attention. A definite place exists in the elementary physical education curriculum for modified team games that involve one or more of the sport skills, rules, or procedures used in official sports such as basketball, softball, and volleyball. Lead-up games present an alternative to the repetitive and often boring skill drills. Such games allow students to go beyond drills of isolated skills in order to play a modification of the game at a level in which success and personal enjoyment go hand in hand. Gallahue and Donnelly (2003) recommend that once the sport skill has been reasonably mastered, skill drills may be modified to take on game form. As students gain proficiency, make the lead-up games increasingly complex by incorporating a greater number of skill elements and/or more involved strategies.

Lead-up games are a way for children to link the simple with the complex. These activities should be viewed as a means to an end because they are preparatory to playing the sport as it is designed to be played. Four examples of lead-up games are presented in figures 7.16, 7.17, 7.18, and 7.19.

Sports.   Sports are classified as (1) individual, such as bowling, golf, and gymnastics; (2) dual, such as badminton, racquetball, and tennis; and (3) team, such as basketball, softball, and volleyball. Official sports are those activities governed by a specific set of rules and regulations and are recognized and interpreted by an official governing body as the standard

Figure 7.16
Sideline Basketball

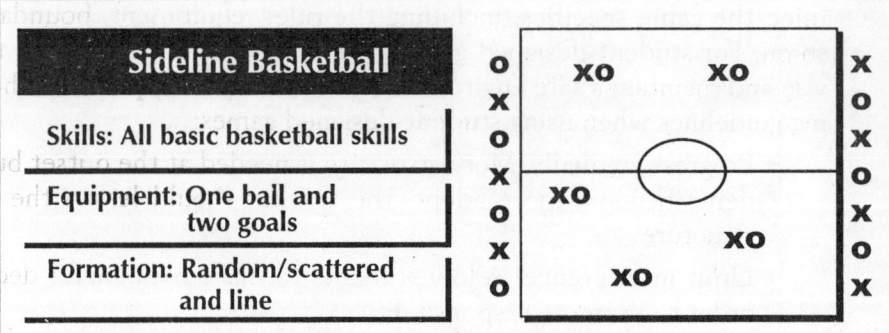

Sideline Basketball
Skills: All basic basketball skills
Equipment: One ball and two goals
Formation: Random/scattered and line

Each team consists of eight to twelve players. Half of the players are on the playing court while the remaining players are alternated by team along each sideline. The object of the game is to move the ball down the court by passing and/or dribbling in order to score a goal. Following a throw-in by either team, standard basketball rules apply. The ball may be passed to a teammate on the sideline who must pass it to a teammate on the court. Following a goal, a defensive rebound, or a stolen ball, the ball must be passed to a sideline player.

Figure 7.17
Zone Soccer

Zone Soccer
Skills: Basic soccer skills
Equipment: One soccer ball per game
Formation: As shown

X_1 O_3 X_1 O_3 X_1 O_3
X_2 O_2 X_2 O_2 X_2 O_2
X_3 O_1 X_3 O_1 X_3 O_1

Each team consists of eight to twelve players. The playing area is separated into thirds with each team divided into three groups and positioned as shown. The object of the game is to kick the ball over the opposing team's goal line. Following a kickoff, players must remain in their respective zones. As a result, the only way to move the ball down the field and attempt to score is by passing it ahead to a teammmate in another zone. X_1 and O_1 players are goalkeepers who are responsible for defending their goal line. X_2 and O_2 players are midfielders who receive passes from the goal keepers and relay them to the forwards (X_3 and O_3) who attempt to score.

Figure 7.18
Straight Base Ball

Straight Base Ball

Skills: Throwing, catching,
and striking

Equipment: Three bases,
one plastic bat, and
two plastic balls

Formation: Random/scattered

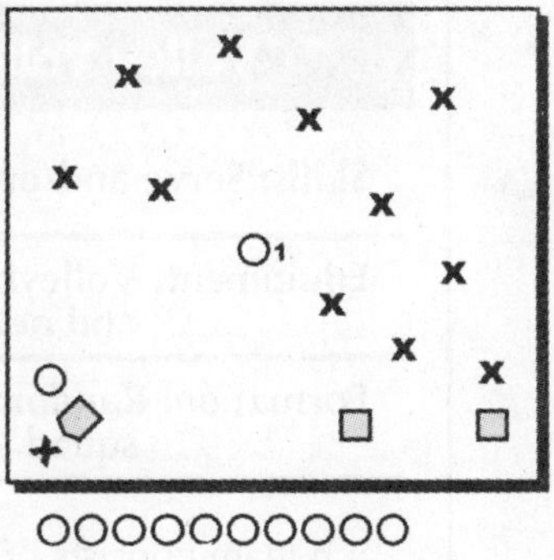

Each team consists of eight to twelve players. The object of the game is to outscore the opponent. The first player on the batting team has only two opportunities to hit a pitched ball. The pitcher is a member of the batting team. After hitting the ball, the batter tries to run to first or second base and return home. The runner cannot stop on a base, and if the player makes it to home safely after touching first base, one run is scored. In addition, if the player touches second base, two runs are scored. The runner is out if (a) a fly ball is caught, or (b) the batter is forced out at home. After each member of the batting team has batted, they take the field.

for performance and play. Many educators agree that official sports are inappropriate for an elementary physical education program. Elementary students do not have the physical/physiological or cognitive development to play official sports as they were designed to be played.

Thoughtful selection and use of games is essential since they have such a potential benefit for the elementary physical education program and because hundreds of games exist from which to choose. For these reasons, Gallahue and Donnelly (2003) provide the following guidelines for today's teacher:

- Choose games that are appropriate for time allotment, class size, space availability, and lesson objective(s).
- Plan in advance so the play area and equipment are ready for use.
- Make necessary explanations clear, concise, and brief with children in game formation.
- Demonstrate or combine the demonstration with the explanation, whenever possible.
- stopping the game too frequently to make corrections.

Figure 7.19
Volley Volleyball

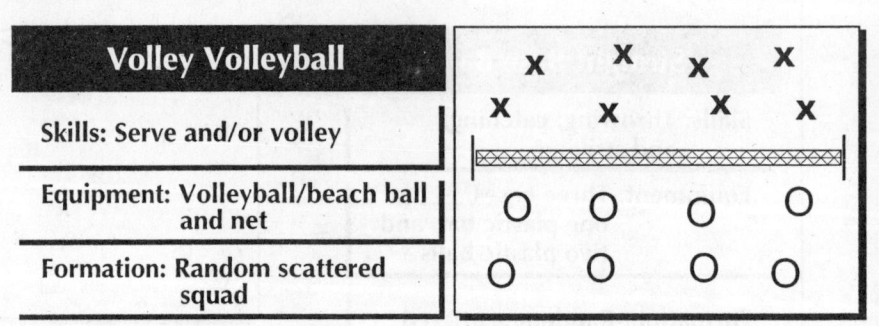

Volley Volleyball	
Skills: Serve and/or volley	
Equipment: Volleyball/beach ball and net	
Formation: Random scattered squad	

Each team consists of six to nine players. The game is played with either a volleyball or a beach ball. The object of the game is to outscore the opponent. The game begins with a serve from anyone on the court. The receiving team may volley the ball as many times as needed to get it back over the net. In addition, any player may hit the ball as many times as necessary. This process is repeated as the teams volley the ball back and forth over the net. The serving team scores one point each time the receiving team (a) fails to return the ball, (b) volleys the ball out of bounds, or (c) touches the net. The game is played to eleven or fifteen points.

Movement Exploration Activities

Read (1945) said that education could be defined as the cultivation of modes of expression—teaching children and adults how to make sounds, images, and movements. Furthermore, the aim of education is the creation of artists—people efficient in these various modes of expression. These insightful statements by Read elevate movement education to new heights. In addition, they encourage both young and old people to ask why humans move, how humans move, and how movement can help them to discover, understand, and adjust to the environment.

Educators agree on what movement exploration can do even though some disagreement exists on both its name and interpretation. Laban (1948) used the term "movement education," stressing the development of a kinesthetic awareness of movements that have both expressive and educational value. Others prefer to call it "movement exploration," partly because it focuses on the discovery of movement potentials. These potentials include movement in a stationary position, movement while in locomotion, movement as a means of communication, and movement as a way to explore one's environment. Others call it "basic movement." Schurr (1980) elaborates on the concept of movement education:

In the primary grades the focus of the physical education program is the exploration and refinement of basic patterns and the understanding of environmental factors that affect movement. Emphasis is placed on the *why* of movement, on understanding one's own capacity in movement, and on acquiring proficiency in a wide range of movement skills. Indirect styles of teaching are utilized to help children develop the processes of exploration, experimentation, simple problem solving, and evaluation. Consistent with the stage of development, children are given a great deal of freedom and responsibility for learning to adapt to elements of intensity; different shapes of small and large objects and obstacles; and in relationship to others as they experience the joy and fulfillment that can accompany movement.

Movement education, therefore, may simply be defined as a means of achieving body management through an understanding of movement factors and the ways they affect the body in motion. While movement exploration is the general term given to the methodology employed for meeting these two objectives, supporters of movement education have made it clear that:

- Individual development of each student is the chief concern.
- Successful movement experiences contribute to an individual's self-confidence and enhance a person's self-image.

Elementary children benefit from varied movement experiences.

- Movements are introduced as age-appropriate challenges.
- Students use personal movements, making them more meaningful.
- Creativity is encouraged, especially in solving selected movement problems.

The movement education concepts are shown in figure 7.20. Body awareness is a knowledge of the various body parts and how these body parts can move. It leads to an understanding of how to move and control the body as a whole or individual body parts in a variety of situations. Spatial awareness is a knowledge of space and an orientation of the body in space, specifically personal, general, and restricted space. It leads to an understanding of the different directional pathways and varying levels in which the body can move. Movement qualities include time, force, and flow. Knowledge of these concepts leads to an understanding of how the body can move from one position to another. Relationships to others and to both small and large apparatus (equipment) occur often in an elementary physical education program. Learning how to move in relation to each of these components is a goal of movement education. These concepts are often presented to students in an exploratory manner, such as when the teacher provides verbal cues in the form of commands, questions, and challenges. Examples of each type of cue follow.

Commands:

- Walk in a circle without touching anyone.
- Run quietly on your toes.
- Skip with a partner while holding hands.
- Jump two times with your hands on your knees.
- Catch the beanbag on the back of your hand.
- Put your elbows together.
- Raise your body from a lower level to a higher level.
- Dribble the ball without using your hands.
- Make your body shake.
- Walk with heavy feet.

Questions:

- Can you walk without swinging your arms?
- Can you move quickly around the play area and stop when you hear the whistle?
- Can you run like an elephant?
- Can you show me how a butterfly moves?
- Can you make believe you are riding in a roller coaster?
- Can you make yourself look like a flower?

• Can you catch the beanbag with one hand?
• Is it possible to kick the ball with your other foot?
• Can you skip with a beanbag on your head?
• How can you catch the beanbag without using your hands?

Challenges:

• Show that you can balance on five body parts.
• Try to move across the play area without walking.
• Try to clap your hands five times before you catch the tossed ball.
• Show how you would walk if it was raining.
• Show that you can catch the ball standing on one foot.
• Show three different ways to move your hand.
• Show that you can dribble the ball with your elbow.
• Show a different way to kick the ball.
• Try to walk across the balance beam backwards.
• Show that you can toss and catch the beanbag from a seated position.

Figure 7.20
Movement Education Concepts

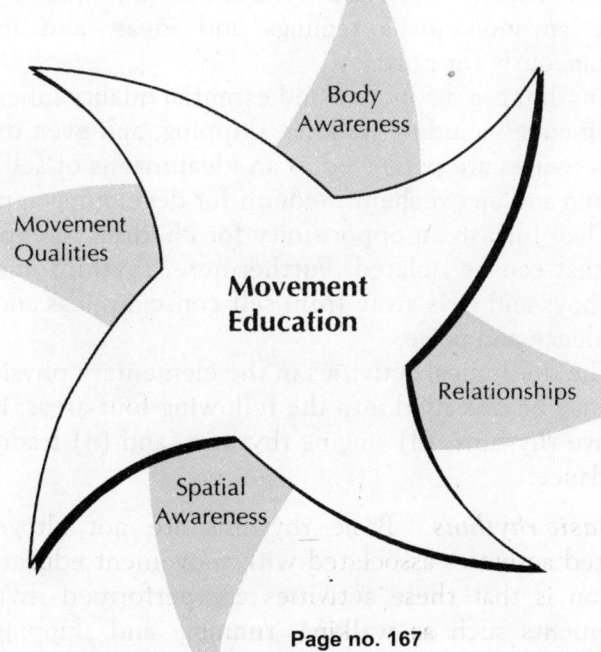

It is important to note that in the process of exploring the fundamental movement skills, whether they are locomotor, nonlocomotor, or manipulative, the student should eventually arrive at a satisfied state of knowing how to execute these skills correctly. For example, a student may have experimented with walking fast, walking slow, or walking with the arms in different positions, but should emerge from these activities with an erect, relaxed, rhythmical walk. In addition, a student may explore how to dribble with the palm of the hand, from a stiff-legged position, or while keeping the ball at eye level, but should eventually dribble using the finger pads, with knees flexed, and with the ball at waist level. From the movement education perspective, it is appropriate to contrast very quick and very slow movements, movements with specific spatial restrictions, and movements from different levels. By using these movement experiences, students learn how to move efficiently and understand the impact of time, force, flow, and space on the body. Furthermore, students can discover the appropriate or skillful way of performing the fundamental movements by experiencing inappropriate ways of doing these skills.

Rhythmic Activities

In virtually all cultures, dance has been a part of the lives of the people, and thus a rich, social inheritance embodied in rhythms and dance exists. Dances have been executed formally as an art form and as a means of self-expression for such events as birth, death, marriage, war, and worship. They have served as a form of recreation, have been used to communicate an individual's feelings and ideas, and have been performed spontaneously for pure joy.

Rhythm is a distinctive and essential quality inherent in all coordinated movements, including walking, skipping, and even dribbling a ball. Rhythmic activities are perceived as an ideal means of self-expression for young children and an excellent medium for developing appropriate social behavior. They furnish an opportunity for children to express almost any emotion that can be isolated. Furthermore, rhythms and dance tend to steer both boys and girls away from self-consciousness and timidity and toward confidence and poise.

The rhythmical activities in the elementary physical education curriculum may be classified into the following four areas: (1) basic rhythms, (2) creative rhythms, (3) singing rhythms, and (4) traditional and contemporary dance.

Basic rhythms. Basic rhythms are not altogether different from selected activities associated with movement education. The primary distinction is that these activities are performed rhythmically. Locomotor movements such as walking, running, and skipping, and nonlocomotor movements such as swinging, swaying, and twisting, can be done to an accompaniment of rhythm sticks, a tambourine, a drum, or an appropriate record, cassette, or CD. As Gallahue and Donnelly (2003) note, a close rhythmic parallel exists between music and movement. With few exceptions, children enjoy both of these activities.

The following examples of basic rhythm activities can be done with students sitting in a random/scattered formation. Using music with a 4/4 meter and a relatively slow tempo, have students listen and:

- Clap on the first beat
- Clap on the first and third beats
- Clap on all four beats
- Slap their thighs on the first beat
- Slap their thighs on the first and third beats
- Slap their thighs on all four beats
- Slap the floor on the first beat
- Slap the floor on the first and third beats
- Slap the floor on all four beats
- Tap their knees on the first beat
- Tap their knees on the first and third beats
- Tap their knees on all four beats
- Do selected combinations of the previous activities: clap their hands four times, then slap their thighs four times; clap their hands on the first and second beats, and slap their thighs on the third and fourth beats; clap their hands on the first and third beats, and slap their thighs on the second and fourth beats; clap their hands on the first beat, slap their thighs on the second beat, slap the floor on the third beat, and tap their knees on the fourth beat.
- Modify the previous activities by having the children perform them as partners facing each other; in groups of three, four, or five facing each other; and while using equipment such as beanbags, paddles, or rhythm sticks.

Creative rhythms. According to Hankin (1992), in a world where discursive language is the primary mode of communication, creative rhythms provide the student an alternative expressive medium. Creative rhythms allow the students to (1) choose their own rhythmical movements, thereby promoting creativity; (2) use a familiar movement vocabulary (fundamental movement skills) rather than having to learn new movements; (3) expand their understanding of the basic rhythms; and (4) express their feelings, ideas, and emotions.

Creative exploration through rhythms may be accomplished by allowing students to perform an exercise routine in time to music, create locomotor and/or nonlocomotor movements to the reading of a poem, or make up a dance routine using music selected by the student. It is important to remember that if the development of creativity is an objective of the elementary physical education program, it should not be left to chance. Instead, the physical educator should encourage curiosity, be respectful of unusual ideas, recognize original creative behavior, ask questions that require thinking, build on skills that students already have, and give opportunities for learning in creative ways.

Singing rhythms. Singing rhythms are activities that enable students to perform prescribed movements to a poem or song while saying or singing the words. Gallahue and Donnelly (2003) indicate that singing rhythms are especially appealing to students because they (1) tell a story, (2) develop an idea, (3) have a pleasing rhythmic pattern, (4) stimulate imagination, and (5) have dramatic possibilities. Furthermore, these simple repetitive rhythms allow the student to learn the words easily and respond through movement while doing so. Examples of singing rhythms include the following:

Did You Ever See a Lassie	London Bridge
Farmer in the Dell	Loobie Loo
Hickory, Dickory, Dock	Mulberry Bush
I'm a Little Teapot	Ring around the Rosie

Traditional and contemporary dance. By definition, this category includes structured dances that require students to perform specific patterns including folk, square, and social dances. Elementary children should be exposed to traditional and contemporary dances because they aid in the development of (1) an understanding and appreciation of the people and cultures of other countries, (2) the ability to transform directions into movements, and (3) desirable social skills and attitudes.

Folk dances are structured dances that are specific to a country. Square dances are specific to North America, reflecting the culture of these countries' forefathers. Social dances reflect the mores of American culture during the times in which they were popular. These dances are done with a partner; therefore, it is recommended that they be introduced in the upper elementary grades. In addition, the interest in contemporary social dances is often short lived, such as disco. The interest in other kinds of dances is cyclic, such as the jitterbug. A listing of traditional and contemporary dances is shown in table 7.5.

Self-Testing Activities

It is not uncommon to hear the modern elementary school criticized because students spend so much time working on teacher-directed activities. This approach is not necessarily harmful, especially if the students become physically involved in the process. What students do, in physical education or any other discipline, should be related to them in a personal way.

Table 7.5
Traditional and Contemporary Dances

Folk	Square	Social
Greensleeves—England	Oh Johnny	Bus Stop
Seven Jumps—Denmark	Birdie in the Cage	Cotton Eye Joe
Virginia Reel—United States	Ladies Chain	Electric Slide
Hora—Israel	Heads and Sides	Fox-trot
La Rospa—Mexico	Red River Valley	Jitterbug
Bleking—Sweden	Dive the Rug	Texas Two Step
Schottishe—Scotland	Duck for the Oyster	Twist
Tinikling—Philippines	Waltz	
Patty Cake Polka—Poland		

Educators should not underestimate the significance of a student's immediate self-appraisal in an activity. Self-testing activities allow students to perform individually; therefore, the individual is given an opportunity to compare (1) personal activity from performance to performance, and (2) peers to himself/herself.

Self-testing activities are important for primary-aged students because they often find it difficult to work cooperatively in group activities. However, these activities also benefit intermediate-aged students by allowing them to learn new skills without having to rely on another student or team as a prerequisite for participation.

Numerous self-test motor activities have great appeal as students strive for recognition. The improvement in an individual's ability to perform selected self-testing activities encourages feelings of adequacy, security, and acceptance. Therefore, setting aside a definite amount of time in the elementary physical education curriculum for self-testing activities is an important practice.

Self-testing activities may be classified in a variety of ways, but they typically include stunts and tumbling activities such as a log roll, a forward roll, and a frog stand; large apparatus activities such as those done on a balance beam, parallel bars, and a horizontal ladder; small apparatus activities using beanbags, hoops, and balls; and additional activities that develop the fundamental movement skills delineated earlier in table 7.1.

Examples of large apparatus activities on a balance beam include having the student walk across the beam:

- On the balls of the feet
- On the tiptoes
- With hands on the head, hips, or knees
- With the body rigid or relaxed
- With the right or left foot leading
- While balancing a beanbag on the head, hand, or shoulder

- While balancing two beanbags on different body parts
- While tossing a beanbag from hand to hand
- While dribbling a basketball on the floor
- While stepping through a hoop

Examples of large apparatus activities using a hoop placed on the floor include having the student:

- Walk/run around the hoop
- Skip/gallop around the hoop
- Stand inside/outside the hoop
- Stand near/far away from the hoop
- Balance on one/two, etc. body parts inside the hoop
- Balance on one/two/etc. body parts inside the hoop and one, two, etc. body parts outside the hoop

Examples of small apparatus activities using beanbags include having the student toss the beanbag into the air and:

- Catch it with both hands
- Catch it with the right or left hand only
- Catch it on the back of the hand
- Catch it on the elbow
- Catch it on the back
- Clap the hands two, three, or more times before catching it
- Count to five, ten, or more before catching it
- Do two, three, or more jumping jacks or jills before catching it
- Say as much of the alphabet as possible before catching it
- Spell designated words before catching it

Examples of small apparatus activities using 9-inch round balloons include having the student:

- Tap it into the air with the right/left hand
- Tap it into the air while alternating hands
- Kick it into the air with the right/left foot
- Kick it into the air with alternating feet
- Tap it into the air using different body parts (elbow, knee, nose, etc.)
- Tap it into the air using two/three/etc. different body parts
- Tap it into the air, drop to the floor in a crabwalk position, and move to catch it between the knees

The self-testing activities selected will depend primarily on class size, the availability of equipment, and the space utilized. However, the age level, ability level, and characteristics of students should be the most important factors when selecting these activities.

TIME ALLOTMENT

After the educator has decided what to teach, the next logical question is how much of it to teach? More specifically, what percentage of time should be allotted to fitness activities, game activities, movement exploration activities, rhythmic activities, and self-testing activities? When progressing from kindergarten to grade six, should more or less time be allotted for a specific content area?

Table 7.6 provides suggested time allotment percentages for the six content areas by grade level. The numbers listed are recommended percentages. Where the emphasis should be placed is dependent not only on the school and nonschool factors discussed in chapter 4, but more importantly on where students range in both skills and knowledge at any grade level. In addition, the movement experiences to which students have already been exposed should be considered. The suggested percentages in table 7.6 are based on the general characteristics and generally accepted needs of elementary school students.

A closer look at the amount of time allocated to the major elements of the elementary physical education curriculum reveals that games and the subsequent development of sport-specific skills grow considerably in importance from the primary to the intermediate level, while movement education activities receive less attention in the intermediate grades. In addition, a small but gradual decrease in emphasis is placed on rhythmic activities from kindergarten to grade six. Interestingly, self-testing activities remain fairly constant in importance across all grades and appeal to all children.

Table 7.6
Time Allotment for Elementary Physical Education Content (in percentages)

Content Area	Grade Level						
	K	1	2	3	4	5	6
Fitness activities	10	10	15	15	20	20	20
Integrated activities	10	15	15	10	5	5	5
Game activities	10	10	10	15	30	30	30
Movement exploration activities	25	20	20	10	10	10	10
Rhythmic activities	25	25	20	20	10	10	10
Self-testing activities	20	20	20	30	25	25	25

SUMMARY

1. Movement is a medium through which students learn about themselves and the world in which they live.

2. Knowledge of the characteristics of elementary school students and the implications arising from these characteristics is essential for developing a physical education curriculum.

3. To meet both the needs and interests of elementary school students as well as the general objectives of the curriculum, a variety of movement experiences is essential.

4. Fitness activities are included in the elementary physical education curriculum to help develop and maintain strength, muscular endurance, flexibility, aerobic endurance, and a favorable body composition.

5. Games are a means of enhancing a student's overall development and well-being. Specifically, games can help students develop the fundamental movement skills and the basic game skills essential for successful participation in sport activities not only during the secondary school years but also throughout adulthood.

6. Movement exploration activities provide a medium for the student to achieve body management through an awareness and understanding of movement concepts and how they affect the body while moving. These concepts include body awareness, spatial awareness, the qualities of movement (time, force, and flow), and relationships.

7. Rhythmic activities provide a means of self-expression and are germane to an elementary physical education curriculum because rhythm is an inherent quality in all movement.

8. Self-testing activities include stunts and tumbling, large apparatus activities, and small apparatus activities.

9. A slowly decreasing percentage of time should be devoted to both movement exploration and rhythm activities from kindergarten to grade six, while an increase should occur for both fitness activities and games. Time for self-testing activities should remain fairly constant from K–6.

QUESTIONS AND LEARNING ACTIVITIES

1. Recess is essential for today's schoolchildren. It affords them an opportunity to be free to play. Children need free play. Do you agree? Is recess sufficient to meet the needs of children?

2. Explain the concept that dance is indeed a basic educational technique.

3. Review a number of sources dealing with movement education. From your reading, formulate your own definition.

4. Find out about play in other cultures. Are games of low organization about the same everywhere?

5. It has been said that the development of motor performance in primary grade students can be advanced more effectively through a

program of specific instruction than through free play and movement exploration. How would you determine whether this is true?
What kind of experiment would you set up?

6. The primary objective for an elementary physical education program
should be the development of health-related fitness. Do you agree or
disagree with this statement? Provide published support for your view.

7. After reading any of Orlick's works, justify a place for cooperative
games in an elementary physical education curriculum.

8. Survey a number of elementary schools to determine the extent to
which games are used in the physical education curriculum. Prepare
a list of the different low-organized and lead-up games used.

REFERENCES

Gabbard, C., LeBlanc, E., & Lowy, S. (1994). *Physical education for children:
Building the foundation.* Englewood Cliffs, NJ: Prentice Hall.

Gallahue, D., & Donnelly, F. (2003). *Developmental physical education for all children.* Champaign, IL: Human Kinetics.

Graham, G. (1977). Helping students design their own games. *Journal of Physical
Education, Recreation, and Dance, 48*(7), 35.

Grant, J. (1995). *Shake, rattle, and learn: Classroom-tested ideas that use movement for active learning.* York, ME: Stenhouse Publishers.

Hankin, T. (1992). Presenting creative dance activities to children: Guidelines for the
non-dancer. *Journal of Physical Education, Recreation, and Dance, 63*(2), 22–24.

Kirchner, G., & Fishburne, G. (1995). *Physical education for elementary school
children.* Dubuque, IA: Brown & Benchmark.

Laban, R. (1948). *Modern educational dance.* London: McDonald & Evans.

Morris, G. (1980). *How to change the games children play.* Minneapolis, MN: Burgess.

Morris, G., & Stiehl, J. (1989). *Changing kids' games.* Champaign, IL: Human
Kinetics.

Orlick, T. (1977). *Winning through cooperation: Competitive insanity.* Washington,
DC: Hawkins & Associates.

Orlick, T. (1978). *The cooperative sports and games book: Challenge with competition.* New York: Pantheon Books.

Orlick, T. (1982). *The second cooperative sports and games book.* New York: Pantheon Books.

Read, H. (1945). *Education through art.* New York: Pantheon Books.

Schurr, E. (1980). *Movement experiences for children.* Englewood Cliffs, NJ: Prentice Hall.

Stillwell, J., & Stockard, J. (1988). *More fitness exercises for children.* Durham,
NC: Great Activities Publishing.

Thomas, K., Lee, A., & Thomas, J. (2000). *Physical education for children: Concepts into practice.* Champaign, IL: Human Kinetics.

Wall, J., & Murray, N. (1994). *Children and movement: Physical education in the
elementary school.* Dubuque, IA: Brown & Benchmark.

Werner, P., & Burton, E. (1979). *Learning through movement.* St. Louis, MO: Mosby.

8

THE SECONDARY PHYSICAL EDUCATION PROGRAM, 7–12

Outcomes

After reading and studying this chapter, you should be able to:
- Define:
    Aquatics
    Ballroom/social dance
    Circuit training
    Conditioning activities
    Folk dance
    Free exercises
    Gymnastics
    Individual/dual activities
    Junior high school
    Contra dance
    Outdoor education
    Senior high school
    Square dance
- Identify the six content areas for secondary physical education programs.
- Describe the physical, intellectual, and social/emotional characteristics of secondary school students.
- Select physical education activities for grades 7–12 based on the students' physical, intellectual, and social/emotional characteristics.
- Discuss the time allotment for each of the physical education content areas for grades 7–12.
- Organize and plan a yearly physical education program for grades 7–12.

P hysical activities can be used to educate and to recreate. Physical educators need to distinguish between these two uses and realize that one precedes the other. That is to say, students need to be *educated* to learn a variety of movement skills that, hopefully, will enable them to *recreate* and lead a physically active life.

It is obvious that a secondary school physical education program must be carefully planned to achieve the goal of meaningful human development. Colleagues require clear direction at the curriculum level to avoid working at cross-purposes or adversely affecting each other's efforts. Therefore, physical educators should examine what is taught under the title of *physical education*. Siedentop (2004) indicates that this is especially important today because the responsibility for education is delegated to the states. As stated in chapter 4, laws governing how much of any subject matter is required within our schools are primarily determined by state legislation. Because states differ dramatically on both their views on education and to what degree they are willing to support this education, it should be no surprise that laws governing the amount of K–12 physical education differ markedly. With today's financial pressures and an increased sense of accountability, it is essential that school districts offer quality physical education programs.

Physical educators and observant general educators are often aware of certain program shortcomings and/or inadequate curriculum content. To some general educators, physical education may appear to be nothing more than aimless play, a criticism that is not without foundation. Inappropriate programs range from disorderly free play to highly organized, inflexible routines. Sport skills and health-related fitness activities may be relegated to the background while games are played period after period as an end in themselves, rather than a means to an end. In such programs, little, if anything, is offered to challenge students. The worst thing about such conditions is that both parents and students can identify a poorly developed curriculum. Today's parents are educated and their secondary students are mature enough to recognize a weak program and/or a weak teacher. During the secondary years, students are either preparing for college or deciding to terminate their education after grade 12. Without knowing which students will continue their formal education, physical educators must assume secondary school students are experiencing their last exposure to a physical education program. A carefully planned and varied curriculum will help ensure that the secondary school physical education experience is a positive and beneficial one for all students.

SECONDARY SCHOOL ORGANIZATION

In terms of pure logic and, to some degree, tradition, it is understandable why the K–6 elementary school years and the 7–12 secondary school years are separated. This logic is used to further separate the junior high school years from the senior high school years. This separation is visible only if the programs are treated distinctly. The inherent problem with this separation is that by studying the two programs in an isolated manner, the significant ingredient of continuity is apt to be ignored or slighted.

If, on the other hand, the physical education curriculum for grades 7–12 is viewed as a continuum, it is possible to (1) appreciate the need for progression in both the selection and presentation of activities, and (2) more effectively observe and assess student progress. Moreover, a number of activities in the junior high school physical education program are repeated in greater depth in the secondary school years. This provides an opportunity to monitor teaching strategies, expectations, and actual outcomes over the span of the secondary school physical education curriculum.

A common format for today's junior high schools is either two years (grades 7 and 8) or three years (grades 7, 8, and 9). These formats are designed to prepare students for four-year (grades 9–12) or three-year (grades 10–12) senior high schools, respectively. In addition, the traditional K–8 grammar school, which directly prepares students for high school, still exists in parts of the country. Even though these are the traditional organizational patterns, there is an increasing number of *middle schools* (grades 6–8). Educators favoring the middle school structure subscribe to the concept that seventh and eighth graders have more in common with upper elementary students than they have with ninth graders.

THE SECONDARY SCHOOL STUDENT

A time occurs in the lives of young people when they are curious and possess an eagerness to discover and try something new. It is a time when *teachable moments* are clearly apparent. The junior high school level is such a time; both boys and girls are capable of extensive physical activity and are willing to struggle, perspire, and concentrate on skill acquisition. To better plan a properly functioning program, curriculum developers need to thoroughly understand the learner's characteristics and the implications arising from these characteristics (see tables 8.1 and 8.2).

The secondary school years should be viewed as a continually evolving period of growth and maturation. The junior high adolescent phase will gradually come to fruition in the later high school years in the form of young adults who are ready to take their place in a college or university environment or in the workforce. Therefore, some of the characteristics of the junior high school student already discussed will apply at the senior high school level, but may have changed in degree. For example, students in grade 10 are more interested in their physical ability and recreational skills than they were previously, yet they still seek peer approval. Throughout the secondary school years, opportunities should be presented to guide and encourage individual and group participation in the types of activities fostering personal security. Ultimately, students should leave school satisfied that their physical education experience has helped them gain self-understanding and has provided them both the desire and the skills to recreate.

Table 8.1
Characteristics of Students Ages 13–15

Physical/Physiological

- Growth is rapid, especially in the long bones of the arms and legs.
- Muscular development is rapid, often resulting in poor coordination or awkwardness.
- Posture is poor.
- Acne period begins.
- Students have a seemingly unlimited source of energy.
- Boys broaden at the shoulders, while girls broaden at hips.
- Voice changes and pubic hair appears in boys as secondary sexual characteristics become apparent.
- Bone ossification is still incomplete.
- In most cases sexual maturity is reached in the later junior high school years.
- Girls are a year and a half to two years ahead of boys in maturation.
- Menstrual cycle is irregular in girls.
- Girls are more concerned with personal appearance.

Intellectual/Social/Emotional

- Students posses a strong desire for independence.
- Students are somewhat rebellious toward those in positions of authority, including parents and teachers.
- Both boys and girls exhibit great interest in group membership and loyalty to group.
- There is intense interest in self-improvement of sport skills, boys more so than girls.
- There is interest in impressing the opposite sex.
- Students move from fantasy to reality, giving way to intellectual analysis.
- There is a keen interest in clothes.
- Boys are more self-conscious of physical inadequacies.
- Students possess a keen interest in competition.
- Students are exuberant, boisterous, and outgoing.
- There is an intense need for friends.

Implications

- Provide both individual and group activities.
- Provide a variety of self-testing activities.
- Provide activities to develop strength and flexibility.
- Employ skill drills to enhance learning and/or improvement.
- Discuss correct technique in skill drills.
- Provide coeducational activities.
- Maintain a safe and well-supervised physical education environment.
- Provide opportunities for the development of responsibility and leadership.
- Provide ample encouragement in all activities.
- Match teams accordingly in competitive

Sources: Darst & Pangrazi, 2002; Gabbard, 1996; Rink, 2004.

Table 8.2
Characteristics of Students Ages 16–18

Physical/Physiological

- Improvement in motor coordination occurs.
- Muscular growth in boys continues as it tapers off in girls.
- Bone growth is nearly complete.
- Maturity in both height and weight is nearly complete.
- Improvement in muscular strength occurs.
- Secondary sex characteristics are complete.
- Increase in endurance occurs.
- Poise, grace, and grooming are more important to girls.

Intellectual/Social/Emotional

- Students are intensely emotional and complex.
- Students are sensitive to limitations in the psychomotor domain.
- Students are well adjusted to the secondary school environment.
- Both boys and girls enjoy exciting, adventurous activities.
- Attempts are made to gain status through social activities.
- Conformity to peer group standards is a dominating influence.
- Students exhibit a keen interest in the opposite sex.
- Students exhibit a growing interest in competition through contemporary sports.
- Students exhibit a strong interest in personal appearance.

Implications

- Continue vigorous physical activity.
- Stress proper form in sport activities.
- Help students learn appropriate social behavior.
- Provide both coeducational and separate activities.
- Provide a cognitive component as to why an active lifestyle is important.
- Survey students in order to provide activities in which they exhibit a keen interest.
- Place a greater emphasis on the quality rather than the quantity of skills learned.

Sources: Darst & Pangrazi, 2002; Gabbard, 1996; Rink, 2004.

PRACTICAL CONSIDERATIONS

Perhaps the foremost concern is knowing exactly what behavioral changes educators want to observe in students following participation in a program. In other words, what are the physical educator's objectives? In previous chapters both program planning and organization were discussed in detail. Earlier chapters emphasized that curriculum content needs to be directly related to specific grade level objectives to produce the best physical education results. Specific objectives need to be highlighted in order to ensure that the curriculum content will lead to student attainment, since both the physical and social/emotional characteristics already listed and the developing needs of today's secondary school students are an issue. These objectives are shown in table 8.3.

Table 8.3
Objectives for Junior and Senior High School

Junior High School	Senior High School
• Development of the physical/ physiological system through vigorous physical activities adapted to the individual	• Continuation of junior high objectives
• Development of motor skills specific to a variety of sports and sport-related activities	• Refinement of these same motor skills as a means of yielding greater satisfaction
• Development of self-confidence, self-direction, initiative, and feelings of personal worth and belonging	• Continued to a greater degree by programming students into extra class activities including intramurals and/or interscholastic athletics
• Development of appropriate social behavior, including proper boy-girl relationships	• Continued to a greater degree through the promotion of co-educational participation
• Development of an understanding of cooperative, democratic living through leadership and followership experiences	• Continuation of junior high objectives
• Development of an understanding of sports and sport-related activities	• Continued to a greater degree to enhance an individual's appreciation of sport activities in order to affect post-school physical education choices

Another secondary school level consideration that is missing at the elementary school level is maintaining a balance between the instructional program and the extra-class program. The instructional program is the portion of the curriculum that provides educational experiences for all students during a regularly scheduled time. The extra-class program is the portion of the curriculum that is conducted beyond the regularly scheduled school day. Examples of extra-class activities are intramurals and interscholastic athletics, which are discussed more thoroughly in chapter 10.

According to Jewett, Bain, and Ennis (1995), the development and implementation of a quality secondary school physical education program requires special attention toward maintaining this balance. To foster this requirement, the following guiding statements should be adhered to while carrying out the instructional program:

• Both boys and girls should be exposed to an instructional program.

• There should be no substitute for the instructional program.

• The instructional program should be scheduled to allow maximum participation, ample time for each student to have an opportunity to be challenged, and opportunities for all students to gain the satisfaction that comes from achievement.

• A variety and significant number of activities should be included in the secondary school physical education curriculum in order to meet all of the objectives.

- A majority of the activities should be introduced at the middle school/junior high school level to acquaint students with the activities. Students will hopefully participate in them through extra-class programs and/or selected community programs.
- The physical educator should allow time for preparation and planning.
- Interscholastic athletics should first and foremost be viewed as an educational experience and secondly as a challenging enrichment program for the physically gifted.

PLANNING AND ORGANIZING THE CONTENT

Three major considerations should be addressed when discussing the mechanics of organizing curriculum content at the secondary school level. The first consideration involves designating the content areas and placing the various activities into appropriate categories; the more commonly accepted content areas for secondary school physical education are shown in figure 8.1. Additional categories include outdoor pursuits such as backpacking, hiking, and orienteering, as well as combative activities such as martial arts and wrestling. A list of nationally taught content is shown in

Figure 8.1
Secondary Physical Education Content Areas

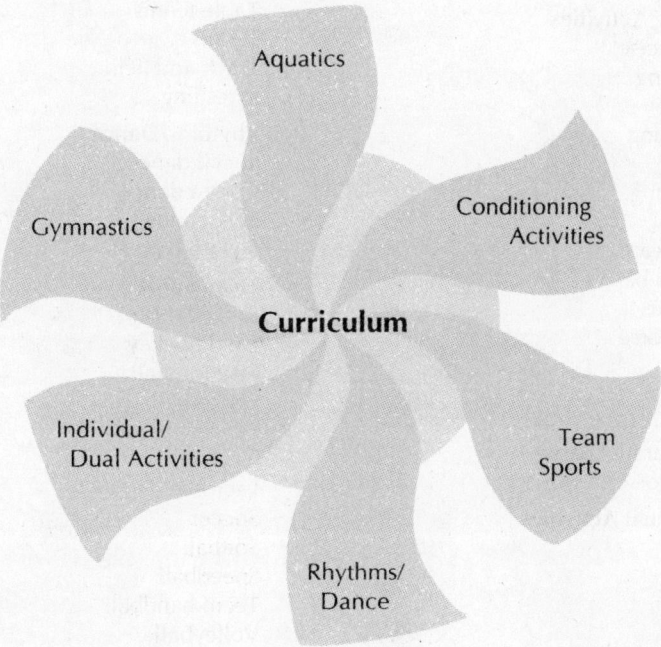

table 8.4. The second consideration pertains to the percentage of time assigned to each of the major content areas; recommended percentages are shown in table 8.5.

The final consideration relates to both the selection of and the percentage of time allotted to the specific activities within each of the six content areas. The instructional framework should be established at this point. In developing a structure for the secondary school physical education program, it is generally advantageous to initially organize the major content areas into blocks of time (units) according to the degree of emphasis considered appropriate for each grade level. Once this development has been accomplished, a further breakdown of time may be determined for the selected activities, which will depend on the number and length of classes per week, the availability of facilities and equipment, and staffing.

Table 8.4
Secondary Physical Education Activities

Aquatics
Diving (basic)
Diving (scuba)
Lifesaving
Snorkeling
Swimming (beginning)
Swimming (intermediate)
Swimming (advanced)
Synchronized swimming
Water polo

Conditioning Activities
Aerobic exercise
Circuit training
Free exercise
Interval training
Jogging
Weight training
Gymnastics
  Balance beam
  Horizontal bar
  Parallel bars
  Pommel horse
  Rings
  Stunts
  Trampoline
  Uneven parallel bars
  Vaulting horse

Individual/Dual Activities
Archery
Badminton
Bowling
Cycling

Individual/Dual Activities (continued)
Deck tennis
Fencing
Golf
Handball
Horseshoes
Racquetball
Shuffleboard
Skating (ice)
Skating (roller)
Table tennis
Tennis
Track and field
Wrestling

Rhythms/Dance
Social dance
Contra dance
Folk dance
Square dance

Team Sports
Basketball
Field hockey
Flag football
Flickerball
Hockey (floor)
Hockey (ice)
Lacrosse
Soccer
Softball
Speedball
Team handball
Volleyball

Table 8.5
Time Allotments for Secondary Physical Education Content (in percentages)

Content Area	Grade Level					
	7	8	9	10	11	12
Aquatics	10	10	10	10	10	10
Conditioning activities	10	10	10	15	15	15
Gymnastics	15	15	15	10	10	10
Individual/dual activities	20	20	20	30	30	30
Rhythms/dance	15	15	15	10	10	10
Team sports	30	30	30	25	25	25

In many schools the selection of activities in each major content area merely tends to follow what other schools have done. The result may be either a mistake or a stroke of genius depending on the other school's guidelines and the soundness of its curriculum development. There is some value in making a list of those schools with exemplary programs at the state level, if not at the national level, and collecting curriculum materials from these schools. It may be advantageous to carefully examine these materials in order to ascertain what they have in common. After determining the major program elements, the physical education faculty should be able to evaluate how well these elements coordinate with the local philosophy and how to use them to best meet local needs.

The validity of a physical education program is often determined by referring to experts who are in a position to judge a program's worth. Such experts may be found in a state's department of education. State supervisors of physical education often have been working in both the design and revision of curriculum for years.

The states of Arkansas, California, North Carolina, Pennsylvania, and South Dakota publish curriculum recommendations. The titles of these frameworks and addresses for obtaining them are listed in table 6.1. Such guidelines provide information relative to

- Developing a philosophy
- Implementing a balanced secondary curriculum
- Sequencing of content
- Assessing the curriculum

A careful review of these resources may suggest ways they can be revised to better suit local needs.

A Closer Look at Content

One reason that curriculum planning is so difficult is that each major content area must be fully developed yet balanced with other areas. In addition, the curriculum must be innovative and flexible. Under such demanding circumstances it would be relatively easy for the administrator merely to select a number of activities from each major content area and then leave the rest to the physical educator. This occurs too often and results in programs that are nearly impossible to defend. A more solid practice would be to structure a somewhat firm and detailed course of study with thorough and balanced coverage of designated content areas. This framework allows some departure by those physical educators who desire to implement innovative activities, but its primary goal is to provide a thorough understanding of the major content areas.

Aquatics

Swimming and water-related activities are essential to a physical education curriculum. Swimming instruction and water survival techniques should be a fixture in public education. In 2001, 859 children ages 1 to 14 died from drowning (Centers for Disease Control and Prevention, 2003). While the drowning rate is slowly declining, it remains the second-leading cause of injury-related death for children ages 1 to 14 years. Aquatic education has been effective in lessening this frequency. Siedentop, Herkowitz, and Rink (1984) provide additional justification for including aquatics in the secondary school program.

1. Swimming is, perhaps, the perfect activity for developing overall health-related fitness. All swimming strokes require use of the limbs and most body joints, resulting in improved strength, muscular endurance, and flexibility. In addition, the energy expenditure associated with swimming helps achieve and maintain an appropriate level of percent body fat.

2. Because such a broad spectrum of swimming activities exists, ranging from basic to more advanced strokes, aquatics can easily provide challenging experiences for all students.

3. Aquatics provide an excellent medium for socialization. In a well-organized aquatics program, students have the opportunity to observe peers, compare their performance with others, help others, and develop leadership skills.

One variable limiting aquatic instruction in today's schools is the lack of a swimming pool, yet a pool on site is not necessary for an aquatics program. Arrangements can often be made to use pools located in public and/or private recreational areas.

Conditioning Activities

A unique goal of physical education is the development and maintenance of health-related fitness—an optimum level of physical condition. This *capacity for activity* has concerned physical educators for centuries. In fact, long before games and recreational activities were accepted as educational activities, the medical profession proclaimed the virtues of exercise (physical activity) as a key contributor to good physical condition and well-being. Since then, a variety of activities specifically designed to promote health-related fitness have been proposed.

What conditioning activities should be included in a secondary school program? The answer must consider both *if* and *how well* the conditioning activities develop the components of health-related fitness. Fortunately, many activities employed for building health-related fitness also develop selected components of motor fitness, including agility, power, and speed. Therefore, it is possible to develop motor fitness skills along with a variety of sport, dance, and game skills while simultaneously developing a gradual but satisfactory level of health-related fitness. This development should not be left entirely to game and sport participation. Other, more specific, health-related fitness activities should be used in the program, such as aerobic exercise, circuit training, free exercise, and weight training.

Aerobic exercise, which leads to an aerobic lifestyle, is a significant curriculum dimension in today's schools. However, the carryover into adulthood is an essential part of aerobic exercise. Aerobic endurance is synonymous with maximal oxygen consumption. This dimension can be measurably improved through repetitive physical activities such as jogging, bicycling, and swimming. This component of health-related fitness is especially valuable during childhood and the adolescent years because people who exercise regularly during the formative years reach adulthood in a state of aerobic fitness, both physiologically and psychologically. Research indicates that those individuals who become physically active at a young age will continue to be physically active in their adult years.

Circuit training is a conditioning activity that enables a large number of students to be active at the same time in a small area, with limited equipment. In addition, circuit training may be easily structured to accommodate students with disabilities since it applies the principle of progressive overload, which involves placing progressively greater-than-normal demands on the muscles being exercised. A circuit of consecutively numbered task stations (eight to twelve, depending on space) is laid out. Students must progress through the circuit and complete the prescribed amount of work at each station. A list of tasks for a junior high school circuit program is shown in table 8.6. Although the basic tasks in the circuit are the same for each student, the intensity at each station will differ depending on the individual student's level of fitness. Therefore, all students can work through the circuit at their own degree of intensity.

Table 8.6
Circuit Program for Junior High School Students

Station	Level 1	Level 2	Level 3
Pull ups	1–3 reps	4–6 reps	Maximum
Run in place	30 strides	40 strides	60 strides
Sit ups	5 reps	11–20 reps	21–30 reps
Bench steps	10–15 reps	16–20 reps	21–25 reps
Push ups	(bent knee) 5 reps	6–10 reps	(hand clap) 6–10 reps
Jump rope	15 reps	25 reps	35 reps
Curls	25 pounds 4 reps	30 pounds 5 reps	35 pounds 8 reps
Pogo springs	30 reps	40 reps	50 reps

*All stations must be completed in 30 seconds.

Free exercises, also called warm-ups and calisthenics, provide a traditional set of movements for enhancing general body conditioning. Called *developmental exercises* by Stillwell and Stockard (1988), these activities are designed to provide proper stimulation to both the musculoskeletal and the cardiorespiratory systems. Since these exercises are designed to be performed in work bouts of either 10 to 25 repetitions or 30- to 60-second intervals, these activities are not generally accepted as *appropriate* for the development of cardiorespiratory endurance. Most of these exercises are categorized as *isotonic* (muscular contraction without significant resistance). In recent years, however, *isometric* exercises (muscular contraction against resistance) have been introduced in the secondary school curriculum. A listing of commonly used developmental exercises is shown in table 8.7.

Weight training has gained popularity with both boys and girls to such an extent that a majority of junior and senior high schools have well-equipped weight rooms. With this increased interest it is important to remember that the emphasis should be placed on overall conditioning and not on weight lifting for competition.

Achieving a comfortable, aerobic lifestyle involves complications—especially in a society that does nearly everything for us. Students who are trying to develop an aerobic lifestyle have to overcome such temptations as sedentary recreation activities (video games), fast foods, and having nearly all physical comforts readily at hand.

Rink (2004) explains that both *incorporating* aerobic fitness and *teaching* aerobic fitness in a school curriculum are difficult for a variety of reasons. One of these difficulties is the insufficient class time allotted for developing fitness, much less maintaining it. Moreover, many older stu-

Table 8.7
Developmental Exercises

Flexibility	Strength/Muscular Endurance
• Elbow knee benders	• Back kicks
• Elbow pullers	• Bent knee sit-ups
• Forward lunges	• Crab walkers
• Head/trunk rotators	• Jump turns
• Hurdle stretchers	• Jump twisters
• Inch worms	• Leg circles
• Needle threaders	• Let downs
• Overhead benders	• Pogo springs
• Side straddle hops	• Push-ups
• Side stretchers	• Reverse push-ups
• Toe touches	• Reverse sit-ups
• Windmills	

dents dislike the work that is necessary for developing/maintaining fitness. Finally, fitness gains are short-lived unless the student maintains a level of activity. To combat these difficulties, Rink suggests the following five curriculum alternatives to assist in achieving the fitness objective:

1. Choose specific grades throughout the curriculum that will focus primarily on fitness. Rather than devoting a small percentage of time to fitness each year, target particular grades (K–12) and strive to meet the health-related fitness objectives.

2. Employ school time outside of the instructional program by implementing fitness activities before school, at lunchtime, after school, and/or in the classroom when appropriate.

3. Approach the development of fitness as a health maintenance behavior by stressing the benefits of leading an aerobic lifestyle. The emphasis is thereby placed on the *why* rather than the *how*.

4. Include activities that have a high fitness value in the curriculum. Therefore, activities with a minimal aerobic component such as archery, bowling, and golf would be excluded.

5. Include some vigorous activity in each lesson. Either modify the lesson content to be more aerobic in nature or incorporate some unrelated fitness activity in the lesson.

Greene (1989) provides examples of such modifications in *Sport Specific Aerobic Routines*; these activities combine aerobic exercise with sport-specific skills. The modification attempts to improve an individual's level of skill while increasing the heart's ability to pump blood more efficiently.

An innovative and varied curriculum will help ensure that the secondary school physical education experience is a positive one for all students.

Greene includes routines for basketball, soccer, softball, tennis, track and field, touch football, and volleyball.

The goal of Rink's five alternatives is to inspire students to not only become fit but to also value fitness. Developing an attitude that will precondition students to be aerobically active is essential.

Gymnastics

Darst and Pangrazi (2002) define gymnastics as specific movements, a performance, or a routine done on either a large mat or large apparatus. Activities done on a mat are called stunts or tumbling when done individually, but they are called floor exercise when grouped together to form a routine. A listing of apparatus and selected stunts is shown in table 8.8. Gymnastic activities

Table 8.8
Secondary School Gymnastic Activities

Apparatus	Stunts
Balance beam	Back walkover
Horizontal bar	Backward roll
Parallel bars	Cartwheel
Pommel horse	Forward roll
Rings	Front walkover
Trampoline	Headstand
Uneven parallel bars	Kip
Vaulting horse	Prone headstand
	Round-off
	Tripod balance

for boys include floor exercises, horizontal bar, parallel bars, pommel horse, rings, trampoline, and vaulting horse, while activities for girls include balance beam, floor exercises, trampoline, uneven parallel bars, and vaulting horse.

Gabbard, LeBlanc, and Lowy (1994) support the inclusion of gymnastics in the curriculum by stating that gymnastics:

- Are inherently challenging
- Are self-testing
- Provide opportunities for creative movement
- Enhance the development of selected health-related fitness components
- Offer a medium for understanding the laws of motion
- Foster cooperation among students

Individual and Dual Activities

A large percentage of time is allotted to games in the upper elementary and early middle school years. These games are generally modified or lead-up games to both team sports and individual/dual activities. The existence of individual/dual activities in the secondary school physical education program is well established. By definition, these activities can be done independently, such as archery, bowling, and golf, or they require either two or four individuals in competition with each other, as in badminton, racquetball, and tennis. Individual/dual activities allow a student to compare his/her performance with other students or with a standard that has been set by a teacher or by the student. At the secondary school level, a generous allotment of time to these activities seems advisable (see table 8.5). The progressive development and perfection of the skills and knowledge involved in these lifetime activities provide a wholesome, enjoyable, physically satisfying form of recreational activity.

Rhythms/Dance

The value of rhythms and dance in elementary education is very important in the physical education curriculum. Much of the previous discussion in this book regarding rhythms and dance tie together the elementary physical education experiences with those at the secondary school level. Activities for grades 7–12 should be an extension of what was experienced in lower grades, in both depth and scope.

Planning a secondary school rhythms and dance program requires physical educators to know what has been offered to students in previous years. It is frequently assumed that students entering junior high school have experienced instruction involving fundamental rhythms that include singing rhythms. In many instances this is not the case. As a result, the physical educator must begin by having students walk, jump, skip, and so on to a variety of audible accompaniments before the various structured dances can be taught.

One of dance's strongest potentials is that it allows students to release human feelings in ways that are significant to the doer, enabling the individual to make a personal statement of what life feels like. It is through the dance medium that students can sort out impressions and sensations by initially internalizing them and then by externalizing them in movement. One of the primary objectives of the National Dance Association is to make dance a part of every individual's experience from early childhood through the retirement years, since it is through this medium that every person has the opportunity to experience self-expression and aesthetic development. In addition, the association's guiding statement explains that every person has the right to move in ways that are expressive, imaginative, primal, and transformational.

The importance of dance in the secondary school curriculum as a means of expression, as an art form, and as a means of physical conditioning is apparent, yet this content area receives little emphasis in many schools. One reason for this lack of emphasis has been the belief that dance is only for girls. This judgmental factor is not only incorrect but should be discouraged since the joy this activity has to offer can and should be shared by both boys and girls. In addition, dance is the ideal coeducational activity for a secondary school physical education curriculum.

The types of dances that may be included in the secondary school curriculum are varied. Different dance types are listed in table 8.4 and discussed in the following sections. Specific dances within each category are shown in table 8.9. The dances that are ultimately selected will be depen-

Table 8.9
Secondary School Dances

Social Dance	Contra Dance	Folk Dance	Square Dance
Cha Cha	Aw Shucks	Admiring the Moon—China	Arkansas Traveler
Charleston	Broken Sixpence	Alexandrovska—Russia	Cowboy Loop
Cotton-eyed Joe	Chorus Jig	Flowers of Edinburgh—Scotland	Four in the Middle
Disco	Easy Does It	Hava Nagila—Israel	Gents Star Right
Electric Slide	Jed's Reel	La Cucaracha—Mexico	Go the Route
Foxtrot	Lady of the Lake	Man in the Hay—Germany	Just a Breeze
Jitterbug	More Again	Sicilian Tarantella—Italy	New Star Thru
Lindy Hop	Shadrack's Delight	Tanko Bushi—Japan	Promenade the Ring
Mambo	Virginia Reel	The Ragg—England	Sail Away
Merengue	Willow Tree	Walczyk "Ges Woda"—Poland	Texas Star
Rumba			
Salsa			
Samba			
Tango			
Waltz			

dent to some degree on the teacher's knowledge of and expertise in the dance; however, this should not be the limiting factor. Individuals are often present within the school and/or the community with the expertise and interest to teach a dance unit.

Social dance. Social dances are structured dances of a recreational nature usually done by couples in a social setting. Pittman, Waller, and Dark (2005) explain that ballroom/social dance has passed through eight periods, each motivated by a specific style of music. These periods include:

- Ragtime, 1900s–1920s
- Tango, 1911–1920s
- Charleston, 1923–1928
- Swing, 1930s–1950s
- Latin, 1930s
- Rock and roll, 1950s
- Twist, 1960s
- Disco, 1970s

The 1980s were characterized by any number of "fad" dance styles, including break dancing, country and western, and hip hop. The dances within the eight periods are somewhat cyclic, as interest comes and goes from time to time. The availability of social settings that offer music (big band, rock and roll, country-western, rhythm and blues, etc.) is limited. Therefore, the need to expose students to social dance in the secondary school physical education curriculum is apparent.

Contra dance. More than 2,000 American contra dances appear in publication today. A contra dance is one that is performed by many couples facing each other in a double-line formation. This type of dance only requires students to be able to count to eight and to move in time with the music. Therefore, contra dances are excellent precursors to the more refined, complex dances.

Folk dance. As defined in chapter 7, folk dances are structured dances that are specific to a country—they are one expression of a country's heritage. Folk dances provide students with both an appreciation for and an understanding of the role that dance has played in the cultural development of various countries. The teacher is encouraged to include cognitive content including the history, geography, and climate of the specific country when teaching the dance. These details will allow students to better understand the customs and traditions of other cultures and appreciate the idea that folk dancing may be a common bond between people of all cultures.

Square dance. Square dance has been proclaimed America's folk dance, dating back to colonial times. Its popularity in the United States is so apparent that a commercial market exists that (1) supports full-time professionals tending to the square dancer's calling and instructional needs, (2) offers square dance vacationing packages, and (3) creates a complete line of fashion wear. Square dance requires students to learn a select number of movements that are fundamental to most dances. Some of these moves include honor (bow), do-si-do, promenade, allemande left, and grand right and left. Once the students learn these moves only one or two additional moves need to be taught in order to learn a new square dance.

The four dance categories just discussed are not all inclusive. Depending on the availability of time and the teacher's expertise, other dances could be taught in the secondary school physical education curriculum. These might include ballet, clogging, jazz, modern dance, round dance, and tap dance.

Team Sports

Sports have become an important element in American culture. As a result, the need for sports education at the secondary school level is also important. Team sports are learned and subsequently played in junior and senior high school for more than just recreation. Team sports are played as a means of developing the quality of *social efficiency*—the ability to get along with others and to exhibit desirable standards of conduct. Abundant evidence exists that validates the positive relationship between motor performance in sport-specific areas and the character traits of leadership and cooperation. During the adolescent years, a student's level of health-related fitness, paired with the development of sports skills, provide an increased capacity for understanding others. This capacity leads to greater social acceptance.

Too often team sports are taken lightly within the instructional program. They are viewed as free-play content, meaning they are played with minimal instruction. This approach results in poorly learned skills because students rush onto courts and fields merely to play. Skills and knowledge must be properly taught if the sport experience is to achieve its objectives. A high concentration of team sports is needed in the secondary school curriculum, with a modest tapering off beginning with grade 10 (see table 8.5). Slowly tapering off allows time for instruction in activities that tend to have a greater degree of carryover into the adult years. Refer to table 8.4 for the most frequently taught team sports. The sports selected as part of the curriculum will be determined partly by the availability of facilities and equipment, geographic location, and, to some degree, student interest.

Outdoor Education

A content area not listed in table 8.4 that has gained popularity in recent years is outdoor education, also called adventure activities or outdoor pursuits. Outdoor education is a means of extending the existing curriculum to enrich the lives of students through outdoor experiences. Specifically, outdoor education is a medium to (1) identify and resolve real life problems, (2) acquire the skills needed to enjoy a lifetime of creative living, and (3) develop a concern about one's natural environment.

Outdoor education includes all pursuits that provide experiences related to the various components of our natural environment—hills, rocks, streams, rivers, trees, and so on. These outdoor pursuits involve adventure, exploration, and a personal contact with nature. Miles and Priest (1990) explain that outdoor education includes structured activities that use natural or artificial environments to identify both individual and group intrapersonal and interpersonal strengths and weaknesses in order to better promote positive personal growth. Specific activities in this content area are listed in table 8.10.

The inclusion of outdoor education activities in the physical education curriculum is one of the most significant innovations in the last hundred years. There is a reluctance to offer outdoor activities because many physical educators (1) are unfamiliar with the content, its purposes, and its benefits, and/or (2) perceive outdoor education as having a different set of educational objectives that require a different set of teaching skills. Yet, including outdoor education activities in the secondary school physical education program allows the physical educator to enhance the relationship between humans and nature. Outdoor pursuits are worthwhile since they provide a medium for developing communication skills, self-concept, self-confidence, cooperation, leadership, and trust. Moreover, adventure activities provide:

Table 8.10
Adventure Activities

Angling	Mountain bicycling
Backpacking	Orienteering
Bicycling	Rappelling
Boating	Rock climbing
Camping	Ropes courses
Canoeing	Sailing
Fishing	Sculling
Fly-casting	Skiing
Kayaking	Sledding

- Active participation regardless of skill level
- Success in challenging activities
- Experience in a different competitive setting, since the student is competing with nature
- A need for responsible behavior, since concern for safety, cooperation, and equipment exists (Siedentop et al., 1984)

If outdoor education is going to be included in the secondary physical education curriculum, the percentage of time devoted to these activities must come from another content area. The choice of which content area to modify should not be made arbitrarily.

SUMMARY

1. In order to develop a meaningful physical education curriculum for secondary school students, an educator must be knowledgeable about the characteristics of these students and the implications arising from these characteristics.

2. At the secondary school level it is essential to maintain a balance between the instructional program and the extra-class program.

3. In order to meet both the needs and interests of secondary school students as well as the general objectives of the curriculum, it is necessary to offer a variety of activities.

4. If facilities are available, then sound justification exists for including aquatics in the secondary school program.

5. Conditioning activities are included in the secondary curriculum to help develop and maintain strength, muscular endurance, flexibility, aerobic endurance, and a favorable body composition.

6. Gymnastics include floor and apparatus activities that challenge students to learn to move in a way different from most sport activities.

7. Individual and dual activities provide a wholesome, enjoyable, physically satisfying means of recreating.

8. Rhythms/dance activities at the secondary school level should be an extension of those experiences included in the K–6 program. The student should have exposure to a variety of dances including ballroom/social, contra, folk, modern, and square.

9. Team sports are an important content area in a secondary physical education curriculum since students can learn a variety of skills for use in their post-school years and appropriate sport behavior (sportsmanship).

QUESTIONS AND LEARNING ACTIVITIES

1. Determine how well Title IX practices are being applied in your area. Where are the strengths? Where are the problems?

2. Visit a school where adventure activities are included in the physical education curriculum. Interview students and physical educators to learn how they value these activities as part of the total physical education curriculum.

3. Review two or three references that discuss the problems of young adolescents. List five or six of these problems and show how physical education may contribute to solutions.

4. Within the profession there are some who feel that the main emphasis of physical education at the secondary level is to teach students

sport skills. Others feel that we should focus on making students fit. Select one view and provide published support for your choice.

5. Survey the recreational facilities in a particular community, both commercial and public. To what extent do the junior and senior high school curricula in physical education provide the appropriate skills, and how do they stimulate interest in the use of these facilities?

6. Is any purpose served by carefully developing an aquatics program on paper if no swimming facilities exist in the community?

7. In your opinion, where does ballroom/social dance fit best in the secondary school physical education program? Support your answer with the viewpoint of at least one reference.

8. Justify the place and importance of a rhythms/dance program at the secondary school level.

REFERENCES

Centers for Disease Control and Prevention. (2003). *Web-based Injury Statistics Query and Reporting System (WISQARS)*. URL: www.cdc.gov/ncipc/wisqars.

Darst, P., & Pangrazi, R. (2002). *Dynamic physical education for secondary school students*, 4th ed. New York: Benjamin Cummings.

Gabbard, C. (1996). *Lifelong motor development*. Dubuque, IA: Brown & Benchmark.

Gabbard, C., LeBlanc, E., & Lowy, S. (1994). *Physical education for children: Building the foundation*. Englewood Cliffs, NJ: Prentice Hall.

Greene, L. (1989). *Sport specific aerobic routines*. Dubuque, IA: Eddie Bowers.

Jewett, A., Bain, L., & Ennis, C. (1995). *The curriculum process in physical education*, 2nd ed. Dubuque, IA: Brown & Benchmark.

Miles, J., & Priest, S. (1990). *Adventure education*. State College, PA: Venture Publishing.

Pittman, A., Waller, M., & Dark, C. (2005). *Dance a while: Handbook for folk, square, and social dance*. San Francisco, CA: Pearson.

Rink, J. (2004). *Teaching physical education for learning*. New York: McGraw-Hill.

Siedentop, D. (2004). *Introduction to physical education, fitness, and sport*. New York: McGraw-Hill.

Siedentop, D., Herkowitz, J., & Rink, J. (1984). *Elementary physical education methods*. Englewood Cliffs, NJ: Prentice Hall.

Stillwell, J., & Stockard, J. (1988). *More fitness exercises for children*. Durham, NC: Great Activities Publishing.

9

ADAPTED PHYSICAL EDUCATION

Outcomes

After reading and studying this chapter, you should be able to:
- Define:

  Adapted physical education

  Disabled

  Exceptional

  Handicapped

  Impaired

  Inclusion

  Mainstreaming

  Special education

- Distinguish among the various organizational structures available that provide for students with disabilities.
- Identify the various components of Public Law 94.142 and discuss their implications for the physical education curriculum.
- Develop and design an IEP.
- Modify and adapt physical education activities to meet a student's individual needs.
- Discuss the individual benefits that can be derived from an adapted physical education program.

P hysically, mentally, socially, and emotionally impaired youth have often been considered exempt from physical education and have thus been unfairly relegated to inactivity in the past. The result is both unrealized potential and social injustice to these youth.

This chapter emphasizes the development of a program that not only concentrates on flexibility and innovation but also is specifically designed to account for individual differences. This kind of program addresses ways to provide active participation for those students who cannot participate safely and successfully in the unrestricted activities of a regular physical education program. This group of students includes more than those with minor ortho-pedic difficulties, postural problems, or low levels of health-related fitness. A comprehensive adapted physical education program will provide opportunities for all students who need individualized and specialized instruction regardless of the disabling condition (see table 9.1). In addition, the program should be a diversified one of developmental movement activities, games, sports, and rhythms suited to the needs and interests, as well as the limitations, of all students.

Table 9.1
Disabling Conditions

Attention deficit/hyperactivity	Convalescing injury
Chronically impaired	Hearing impaired
Anemia	Learning disabled
Arthritis	Mentally retarded
Asthma	Mild
Autism	Severe
Cardiac disorders	Multi-handicapped
Cerebral palsy	Neurologically impaired
Diabetes	Obese
Epilepsy	Orthopedically impaired
Hemophilia	Overweight
Multiple sclerosis	Speech impaired
Muscular dystrophy	Terminally ill
Poliomyelitis	Visually impaired
Spina bifida	

DEFINITIONS

In order to better understand the need for and importance of adapted physical education programs, it is useful to first define some key terms.

- *Adapted physical education.* An individualized program of physical and motor fitness; fundamental motor skills and patterns; and skills in aquatics, dance, and individual and group games and sports designed to meet the unique physical education needs of individuals (Winnick, 2000).

- *Adapted physical educator.* A professional with specialized training in designing, implementing, and evaluating specialized physical education programs (Auxter, Pyfer, & Huettig, 2004).

- *Disabled*. Individuals who have lost physical, social, or psychological functioning that significantly interferes with normal growth and development (Hardman, Drew, & Egan, 1987).

- *Exceptional*. Individuals so different in mental, physical, emotional, or behavioral characteristics that in the interest of equality of educational opportunity, special provisions must be made for their proper education (Daniels & Davies, 1982).

- *Handicapped*. A limitation that is imposed on the individual by environmental demands and that is related to the individual's ability to adapt to these environmental demands (Hardman et al., 1987).

- *Impaired*. Individuals with an identifiable organic or functional condition that adversely affects their educational performance (Seaman & DePauw, 1989).

- *Inclusion*. Educating students with disabilities in regular educational settings with nondisabled students (Winnick, 2000).

- *Mainstreaming*. Placement of handicapped students into the regular physical education class with an Individualized Education Program (Auxter et al., 2004).

- *Special education*. Specially designed instruction, at no cost to the parents or guardians, to meet the unique needs of a handicapped student including classroom instruction, instruction in physical education, home instruction, and instruction in hospitals and institutions (U.S. Congress, Public Law 94.142, 1975).

BENEFITS OF ADAPTED PHYSICAL EDUCATION

Before developing an adapted physical education program, it is important to answer the following questions:

- What behavioral changes do educators want to see in students with disabilities?

- What accomplishments may be expected from students as a result of being involved in the program?

- What individual benefits may be derived from such a program?

The benefits that may be derived from participation in an adapted physical education program include the development of:

- Health-related fitness
- An understanding of one's movement capabilities and limitations
- Physical capacity and joint function
- A variety of recreational skills

- An understanding, appreciation, and enjoyment of movement
- Both satisfaction and a sense of pride in one's ability to overcome limitations
- A positive self-image
- Effective interpersonal relationships

THE MANDATE FOR ADAPTED PHYSICAL EDUCATION

Theoretically, physical education offers something to every student. Merely excusing disabled students—individuals unable to perform in the regular physical education curriculum—is to sidestep responsibility. What must be provided is a physical education curriculum that specifically allows for individual differences.

The need for adapted physical education has never been greater than it is today. In 2000–2001 more than 6 million students with a variety of disabilities were served under Public Law 101.476, Individuals with Disabilities Education Act (IDEA). Adapted physical education, as we know it

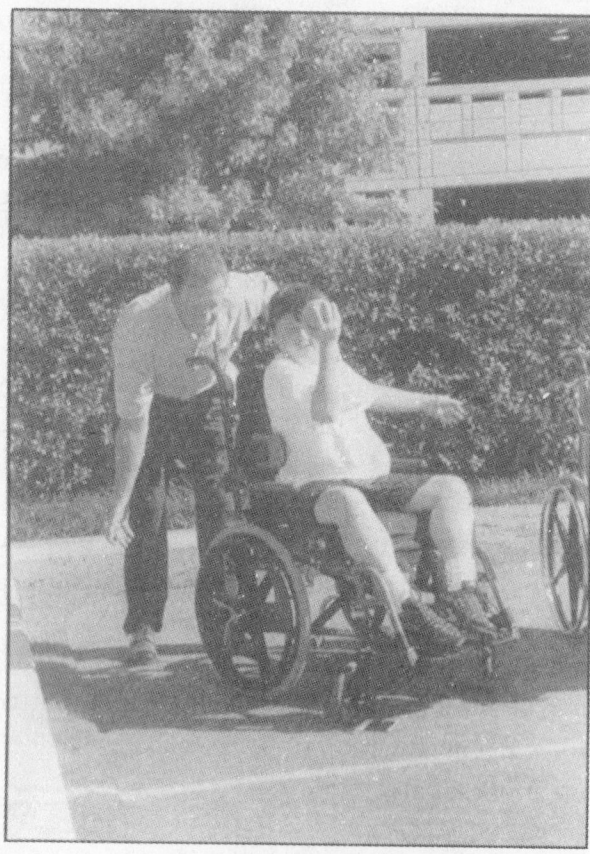

Physical education has something to offer to all students.

today, began as a result of the passage of Public Law 93.112 in 1973. Since that time, continuous governmental support for the education of persons with disabilities has occurred (see table 9.2).

The passage of these legislative acts has direct implications not only for the physical education curriculum, but also for the physical education instructor. In order to both help disabled students and to operate within the federal mandate, today's physical educator needs special knowledge of the various disabling conditions; the laws affecting teachers; motor, perceptual-motor, and fitness tests for diagnostic assessment; and individualized education programs.

Public Law 93.112

In 1973 Congress passed Public Law 93.112, also known as the Rehabilitation Act. Section 504 of this act guaranteed the civil and personal rights of handicapped individuals in all programs for which the sponsoring group(s) received federal funds. The responsibility falls on local education agencies—working through individual schools—to see that appropriate and necessary accommodations are made and individuals are not withheld from, discriminated against, or excluded from activities due to a handicapping condition.

Table 9.2
Federal Legislation Impacting Individuals with Disabilities

Name	Focus
1973 *Public Law 93.112*, Section 504 Rehabilitation Act	Declared that handicapped people cannot be excluded from any program or activity receiving federal funds on the sole basis of being handicapped.
1975 *Public Law 94.142* Education for All Handicapped Children Act	Required that an individual education program be developed for each handicapped child and that handicapped students receive their programs in the least restrictive environment.
1983 *Public Law 98.199* Amendments to the Education for All Handicapped Children Act	States were required to collect data to determine the anticipated service needs for handicapped children.
1986 *Public Law 99.457* Education for All Handicapped Children Amendment	States were instructed to develop comprehensive interdisciplinary services for handicapped children from birth to age two and to expand services for handicapped children ranging in ages from three to five.
1990 *Public Law 101.476* Individuals with Disabilities Education Act	Replaced and extended provisions of Education for all Handicapped Children Act to further guarantee the educational rights of students with disabilities.
1997 *Public Law 105.17* Individuals with Disabilities Education Act Amendments	Provided changes in the law, including provisions for free, appropriate education for all disabled children ages 3–21.

Public Law 94.142

The rules pertaining to the implementation of the Rehabilitation Act were enforced in December 1976 as a part of Public Law 94.142, also called the Education for All Handicapped Children Act, which was signed into law by President Gerald Ford on November 29, 1975. This landmark law is currently enacted as the Individuals with Disabilities Education Act (IDEA), as amended in 1997. At that time the act was designed to ensure that all handicapped students had access to a free, appropriate public education including special education and related services for meeting their unique needs. In addition, the law (1) ensured that the rights of handicapped students and their parents would be protected; (2) gave assistance to states and localities that provided for the education of all handicapped students; and (3) required ongoing assessment to ensure the effectiveness of efforts for educating these students.

The provisions of this law were expected to be fully implemented in school curriculums by 1980. The following procedures were established to make sure that adopted programs address students' needs.

1. Educators are to conduct a needs assessment program that establishes realistic goals for each learner. Individual testing will determine the extent to which the individual attains the goals. The gaps between the established goals and the level of the learner will be identified and put in statement form.

2. Educators will plan and initiate the Individualized Education Program (IEP).

3. Educators will evaluate learner progress at least once annually.

4. Each learner will be placed in the least restrictive environment; that is, the setting in which the student is likely to have the most productive learning experiences (see figure 9.1).

5. Parents or guardians will be advised at periodic intervals as to the progress of the student.

For the purposes of this act, physical education includes special physical education, adapted physical education, movement education, and motor development. It entails the development of health-related and motor fitness, fundamental motor skills, body mechanics, individual and group games, and sport skills such as intramural lifetime sports, aquatics, and dance. In addition, each disabled student must be given the opportunity to participate in the regular physical education program that is available to nondisabled students, unless the student is enrolled full-time in a special facility. If specially designed physical education is prescribed, then the public agency responsible for this education must provide services directly or make arrangements for them through other public or private programs.

Figure 9.1
Adapted Physical Education Placements

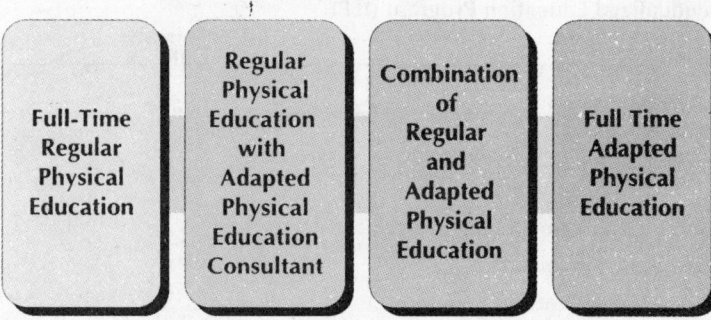

INDIVIDUALIZED EDUCATION PROGRAMS

In Section 602 of the Education for All Handicapped Children Act, considerable attention is directed to the process of designing instruction and activities that meet the unique needs of students through an Individualized Education Program (IEP). Carefully defined in the regulations, the IEP must include a written statement for each handicapped student including annual goals, short-term objectives, and services to be provided (see figure 9.2). The U.S. Department of Education's Office of Special Education and Rehabilitative Services (2005) indicates that the IEP serves as a:

- Focal point for parent-school communication and a tool for resolving differences between the two parties
- Basis for evaluating the extent of the student's educational progress
- Monitoring device for governmental agencies attempting to determine if a student is truly receiving a free, appropriate education
- Source for listing all services that must be provided to the student and, as such, is used to determine the appropriate placement of, as well as the appropriate curriculum for, the student

Prior to the start of the student's program, six IEP procedures take place. First, educators must receive a formal parental consent and preplan the student's objectives. Next they assess the student and write a specific description of the disabling condition(s). A mandatory conference between the parents and the teachers responsible for the student's education is the next step. Participants will discuss the IEP as it pertains to the student's level of educational performance—instructional goals and objectives, special education, related services, and projected time schedules. Following this conference, the parties will discuss the services available for

Figure 9.2
Worksheet for Developing an IEP

Individualized Education Program (IEP)

Student: _____ Date: _____

School: _____ Grade: _____ Age: _____

a. Performance level assessment

b. Annual student goals

c. Short-term instructional objectives

d. Educator(s) involved

   • Special education

   • Physical education

e. Program starting date

f. Program ending date

g. Date of review procedure

h. Program particulars:

   • Placement

   • Curriculum

   • Time assigned to inclusive physical education class

   • Time assigned to adapted physical education

i. Committee recommendations: learning styles, teaching materials, etc.

j. Evaluation criteria

k. Members of the IEP committee

l. Committee meeting dates

meeting the needs identified from the assessment. The fifth step is program implementation, a carefully written statement that is disseminated to all personnel participating in the student's IEP. Finally, the student's progress is evaluated, as well as the suitability of the existing program and instructional objectives.

COOPERATIVE PLANNING AND COORDINATION OF PROFESSIONALS

Any effective program that meets the legislative requirements for disabled learners must involve a close working arrangement between all individuals responsible for instruction. The need for a cooperative effort involving the school health personnel, classroom teacher, special education teacher, physical educator, family physician, and parents is essential to ensure that all students with disabilities receive the appropriate physical education. Cooperative curriculum planning can make it possible to provide appropriate and challenging movement experiences involving individual, partner, small-group, and large-group activities. The challenge for the professionals who deliver services to students with disabilities is considerable and requires cooperation if the objectives are to be met.

Individuals who provide services outside the school must also coordinate their activities with school personnel. In addition to the direct medical services provided for the disabled student by physicians and nurses, psychological services are frequently provided by a counselor or a clinical social worker. Providing a specific rehabilitation program with the goal of returning the disabled student to an effective level of overall well-being requires a number of additional personnel, including those engaged in physical, occupational, and recreational therapy. Without adequate coordination, problems may arise in the delivery of the related services, including duplication of efforts among the many professionals involved (Auxter et al., 2001).

Although a wider variety of students today require the attention of the adapted physical educator, the traditional tasks still remain a part of an effective program. Such tasks include working with children who have low levels of fitness and postural problems, guiding the slow learner to gain general motor skills, and planning programs for the obese. For the most part, these students are not considered special education students, as defined in the Education for All Handicapped Children Act. Most of these individuals will be students who have no restrictions relating to their ability to move, yet attention to their needs is just as important as to the needs of students who require special education under the act.

CONSULTATION IN ADAPTED PHYSICAL EDUCATION

Block, Brodeur, and Brady (2001) indicate that the role of the adapted physical education (APE) specialist has changed since the 1990s. Until recent times the primary responsibility was providing direct services to students with disabilities. But due to inclusion and the ever-increasing number of disabled students, consulting has become one of their more important responsibilities.

In this role the APE specialist provides consulting to physical educators and classroom teachers who then provide direct services to the students with disabilities. APE consultants typically offer information about specific disabilities, safety concerns, activity modifications, methodology specific to the disability, the IEP process, assessment, and advocacy (Auxter et al., 2001; Block & Conatser, 1999).

CLASSIFICATION AND ORGANIZATION

As previously stated, children with disabilities should be placed in a physical education class that will provide the most effective learning experiences. Working together, the special education teacher and physical educator should classify and place the students accordingly. The assessment techniques used must be valid to ensure proper placement. The student's family physician should be consulted since this health professional has the medical information necessary to help determine appropriate physical activity. This meeting will allow the physical educator an opportunity to share with the physician both the goals of the adapted program and the commitment to providing quality physical education for the disabled student. To meet this objective, it is recommended that the family physician complete a medical referral form. Figures 9.3 and 9.4 are examples of departmental letters dealing with activity expectations to be completed by the attending physician.

A variety of organizational structures are available to provide services for students with disabilities. These range from the informal to the more formal and from the small, homogeneous grouping to the larger, heterogeneous grouping. This is partly due to the fact that no single structure can provide for the various disabling conditions. One model for grouping disabled students for instruction is to use the least restrictive environment approach previously discussed in this chapter. In this model, each student is assessed and then receives the necessary instruction in one of the four categories shown in figure 9.1. A second, more practical model involves placing the student in one or more of the following categories:

- *Unrestricted activity.* This involves total inclusion, allowing for full student participation in all physical education activities.

- *Restricted activity.* This category calls for partial inclusion, limiting student participation to teacher-designated activities.

- *Adapted activity.* This allows for participation in modified activities based on the physical, mental, and emotional limitations of the student. Modifications may be made specific to the activity being considered (see table 9.3) or specific to the participant(s) involved in the activity (see table 9.4). There are a variety of ways to modify an activity, and these modifications should be planned carefully.

- *Remedial activity.* Students participate in teacher prescribed exercises and movement activities for correcting defective body mechanics and perceptual-motor functioning.

- *Developmental activity.* This program contains the elements of both the adapted and the remedial programs. It is comprised of movement activities designed to improve basic skills, physical fitness, and socialization.

Figure 9.3
Student with Disability Activity Referral Form

Physician's Recommendation for Physical Education

Dear Physician:

All students enrolled in public schools participate in physical education activities that are designed to meet their growth and developmental needs. In addition, many students participate in other types of physical activities, including intramurals, athletics, etc. To identify the specific needs of students, all school personnel, parents, and the physician must work cooperatively. Please provide the information below so that the most appropriate activities may be provided.

I have examined_____ and find the following disability(es):

_____

_____

_____

I recommend the following (please check all that apply):

_____ 1. No activity restrictions.

_____ 2. Participation in all activities except athletics.

_____ 3. No restrictions in physical education.

_____ 4. Adaptations in physical education:

      _____ a. Limited running and jumping

      _____ b. No running and jumping

      _____ c. No activities involving body contact

      _____ d. No strenuous conditioning exercises

      _____ e. Exercises designed for rehabilitation only

_____ 5. Other adaptations (please specify):

_____

_____

I recommend these adaptations for:

_____ 2 weeks _____ 2 months

_____ 1 month _____ Other (please specify)

Date: _____

Physician: _____ Address: _____

Please return this form to:

_____

_____

_____

(School name and address)

Figure 9.4
Adapted Physical Education Referral Form

Department of Physical Education

Name: _____

Date: _____.

The physical education department at ABC High School believes that every student who is able to attend should be able to benefit from the program. After allowing time for undressing, dressing, and showering, the daily activity time ranges from twenty-five to thirty minutes.

Please either (X) one general activity category or (X) specific activities you would recommend for this student.

Mild	Moderate	Strenuous
( ) Aerobic exercise	( ) Aerobic exercise	( ) Aerobic exercise
( ) Archery	( ) Apparatus (bars)	( ) Basketball
( ) Corrective exercise	( ) Apparatus (horse)	( ) Cageball
( ) Golf skills	( ) Badminton	( ) Field hockey
( ) Horseshoes	( ) Badminton skills	( ) Soccer
( ) Mild dance	( ) Bowling	( ) Speedball
( ) Shuffleboard	( ) Corrective exercise	
( ) Table tennis	( ) Dance	
( ) Tennis skills	( ) Golf	
( ) Tumbling	( ) Softball	
	( ) Tennis	
	( ) Volleyball	
	( ) Volleyball skills	
	( ) Weight training	

Note: If strenuous activity is recommended, it is accepted that mild and moderate are permissible unless exceptions are indicated. Likewise, if moderate activity is recommended, it is accepted that mild is permissible. If you think the student should be doing only mild activities, but find that one or more moderate activities such as bowling and golf should be included, then check them. If non-activity is desirable at this time for this student, please make this notation and provide the number of days this status should continue.

Up to the date of _____ , please excuse this student from all activity.

Signed _____
                    Attending Physician

Comments: _____

_____

_____

These categories are suitable not only for the identifiably disabled student but also for students who (1) are convalescing from an illness or injury that necessitates a gradual return to unrestricted activity; (2) need to strengthen certain muscle groups; and (3) have postural deviations that can be improved through prescribed activity.

Table 9.3
Activity Modifications

1. Reduce playing area
2. Use larger equipment
3. Use lighter equipment
4. Shorten playing time
5. Reduce point requirements for winning
6. Eliminate competition by not keeping score
7. Allow balls to bounce or be caught in games like volleyball
8. Use special devices to accommodate (handrails, guide ropes, and textured floors)
9. Lower baskets and nets
10. Decrease number of repetitions in exercises
11. Reduce tempo in rhythmic activities
12. Increase size of striking implements and targets
13. Use stationary ball-handling activities
14. Stress process (technique) rather than product (performance)
15. Change the formation

Table 9.4
Participant Modifications

1. Use extra players
2. Use free substitutions
3. Rotate players frequently
4. Allow frequent rest periods
5. Allow two hands instead of one when accuracy/power is involved
6. Allow substitute runners
7. Substitute walking for running
8. Eliminate movements of specific body parts
9. Take fewer turns
10. Restrict playing area to half court or less
11. Play a team position requiring less activity
12. Allow peer assistance

INCLUSION

By law, students with disabilities cannot be excluded from physical education. In the past, programming for these students has typically taken one of the following forms: (1) a segregated special education class, or (2) an integrated class.

The enactment of Public Law 94.142 in 1975 resulted in a transition from separate special education programming to inclusive programming. Kirchner and Fishburne (1995) indicate that two additional factors also contributed to this change. First, today's parents are more aware of their children's educational rights. Second, dwindling resources have forced state administrators to reduce monetary support for special programs. As a result, public schools must assume more of the responsibility for providing both the programs and the staff to address the educational needs of the special student.

Whenever possible, it is recommended that students with disabilities be integrated—placed in their regular physical education class. Originally called *mainstreaming* in the 1970s, it provided the foundation for what is now called *inclusion*. Mainstreaming focused on placing the student in the least restrictive environment, whereas inclusion focuses on the learned outcomes achieved by the student. It is based on the belief that children who learn together will learn to live together. The inclusion movement has gained support from many educators who believe that a separate education is not an equal education (Winnick, 2000).

Inclusion suggests that students with disabilities receive their IEP in physical education in the context of general physical education, with the necessary adaptations and support to ensure appropriateness, safety, and success (Block, 1994). Inclusion allows all students, regardless of disabling condition, to be included in every aspect of school life in an attempt to allow them to be a true part of their school community. Inclusion has become widespread because of the following benefits:

- Age-appropriate role models are available
- Students have the opportunity to learn appropriate social skills
- Tolerance and acceptance are encouraged
- The stigma of separation and labeling is lessened

To better meet the needs of students in an inclusive physical education environment, Rizzo and Lavay (2000) state that educators must consider such factors as the student's skill level, motivation, and need for behavior management, and must also be prepared to make the appropriate modifications in methodology and curriculum (see table 9.5). Block and Vogler (1994) discuss three changes that have proven useful when curricular modification is warranted:

Table 9.5
Inclusion Factors and Implications

Factor	Implication(s)
Skill level	Provide a variety of activities to accommodate students with a broad range of skills.
Motivation	Provide a positive environment. Praise often, but genuinely.
Methodology	Utilize a wide range of instructional techniques as needed to be effective. This includes individualized instruction, self-paced work, partner work, small group/large group activity, and peer teaching (see chapter 6).
Behavior Management	Determine which students need assistance following class rules. A specially designed behavior management program may be needed for each student.
Modifications	Make necessary curricula modifications to allow for student success.

Source: Adapted from Rizzo & Lavay, 2000.

1. Changing the content or *what* is taught.

2. Changing the methodology or *how* it is taught.

3. Changing *who* does the teaching.

Changing the content or *what is taught* may be accomplished in either a multilevel or overlapping format. In the multilevel format, all students are involved in the same unit (basketball) and the same skill (dribbling, passing, or shooting) within that unit. However, the learning experiences used are selected in relation to each student's abilities. For example, students without disabilities may be working at the application level by using both complex game skills and strategies in a 3-on-3, game-like drill. In the same class, disabled students may be working at the comprehension level by learning basic skills and a lower-level strategy in a 2-on-1, game-like drill. In the overlapping format, students work toward meeting the same goal. In other words, a shooting objective for all students in a basketball unit may be to make three out of five lay-ups. To ensure that disabled students can attain this goal, the basket may need to be lowered and/or the size of the basketball may need to be changed.

Changing *how the content is taught* involves activity modifications to suit individual needs. This format is based on the game/activity modification work of Morris and Stiehl (1989). Existing sports such as basketball, soccer, and volleyball are modified; lead-up games such as sideline basketball, zone soccer, and volley volleyball can be further modified to allow all students to participate successfully.

Changing *who will teach the content* means that the educational services provided by the physical educator are supplemented with those of additional individuals who are employed to better meet the needs of disabled students. Adapted physical education specialists, paraprofessionals, parents, and peers are among the people providing these services. Craft (1996) states that it is important to remember that inclusion changes the assumption that the physical educator is employed to deliver standard curriculum content to students. Instead, the educator is employed to teach content that is appropriate for each student, which may vary greatly among students. Helion and Fry (1995) indicate that when modifying an activity, the teacher may take two approaches. The physical educator may manipulate either the task or activity itself or the environment in which the task is being done (see tables 9.3 and 9.4 for specific modifications).

SCHEDULING

The scheduling of special classes for students with disabilities depends not only on the nature of the disabling condition but also on the availability of both a teaching station (gymnasium, classroom, playground, and so on) and an instructor. When the intent is clearly to help a student in need of individual attention, the means will be found. A few students can be easily accommodated during or after school hours, but if numbers are large, problems arise. If faculty are available and at least two teachers are assigned to a class, one should be able to work primarily with the disabled students who need special attention. If only one physical educator is available, the use of student leaders may allow the teacher to direct some attention toward students who can profit from individual attention.

Even though noon hour, after school, and study periods are times that have been used for special scheduling, the most efficient programs are preplanned and taught during the regular school day. Scheduling a separate adapted physical education class at the same time as a regular class is an acceptable method. For disabled students who need additional services but are placed in an inclusion curriculum, additional classes may be scheduled (1) on alternate days from the regular class; (2) during alternate periods; or (3) during elective periods. To avoid disrupting the established school schedule, it is advantageous to assess the disabled students in May before school is dismissed for the summer. At this time, a determination can be made as to which students will profit most from an adapted physical education program that would begin the following August. This allows ample time to develop a workable schedule and answer questions relating to faculty, class size, and teaching stations.

SUMMARY

1. To better meet the needs of students with disabilities, physical education activities often need to be adapted.

2. Teaching in an inclusive physical education setting requires that teachers promote acceptance of all students, modify the curriculum to better meet students' needs, and work collaboratively with all involved.

3. Public Law 94.142 ensures that all disabled students be provided a free, appropriate education and the related services to meet their needs. To meet these needs, an Individualized Education Program (IEP) must be developed for each student.

4. More than one individual is often involved in organizing, implementing, and evaluating the IEP for the disabled student. For the best results, this should be a collaborative and cooperative effort including the physical educator, classroom teacher, physical therapist, family physician, and parents.

5. A variety of organizational structures are available to deliver physical education to disabled students. The structure selected should provide for the most effective learning.

6. Inclusion has become the accepted method of programming adapted physical education.

QUESTIONS AND LEARNING ACTIVITIES

1. Prepare a statement that will support the premise that a proper program of physical education will provide an opportunity for the development of physical expression to counteract the lack of verbal expression in mentally disabled students.

2. What is the difference between an adapted program and a remedial program?

3. What are the obstacles to overcome in establishing an adapted physical education curriculum in a school system that has never had this kind of program?

4. Visit a small school and a large school to examine the quality of their physical education programs for students with disabilities. Find out their objectives, content, and evaluative techniques.

5. Interview several teachers of special education. Determine their views relating to the value of physical education as compared to a free-play period for students with disabilities.

6. Conduct research and prepare a description of an existing innovative program for disabled students.

7. List what you think are the three major advantages and disadvantages of inclusion programming. Provide published support for your answer.

REFERENCES

Auxter, D., Pyfer, J., & Huettig, C. (2004). *Principles and methods of adapted physical education and recreation*, 10th ed. New York: McGraw-Hill.

Block, M. (1994). Why all students with disabilities should be included in regular physical education. *Palaestra*, *10*(3), 17–24.

Block, M., & Conatser, P. (1999). Consulting in adapted physical education. *Adapted Physical Activity Quarterly, 16*, 9–26.

Block, M., Brodeur, S., & Brady, W. (2001). Planning and documenting consultation in adapted physical education. *Journal of Physical Education, Recreation, and Dance, 72*(8), 49-52.

Block, M., & Vogler, E. (1994). Innovative and adaptive curriculum models for full inclusion. *Teaching Elementary Physical Education, 5*(5), 6–7.

Craft, D. (1996). A focus on inclusion in physical education. In *Physical education sourcebook*. Champaign, IL: Human Kinetics.

Daniels, A., & Davies, E. (1982). *Adapted physical education*. New York: Harper & Row.

Hardman, M., Drew, C., & Egan, M. (1987). *Human exceptionality, society, school and family*. Boston: Allyn & Bacon.

Helion, J., & Fry, F. (1995). Modifying activities for developmental appropriateness. *Journal of Physical Education, Recreation, and Dance, 66*(7), 57–59.

Kirchner, G., & Fishburne, G. (1995). *Physical education for elementary school children*. Dubuque, IA: Brown & Benchmark.

Morris, G., & Stiehl, J. (1989). *Changing kids' games*. Champaign, IL: Human Kinetics.

Rizzo, T., & Lavay, B. (2000). Inclusion: Why the confusion? *Journal of Physical Education, Recreation, and Dance, 71*(4), 32–36.

Seaman, J., & DePauw, K. (1989). *The new adapted physical education: A developmental approach*. Mountain View, CA: Mayfield.

U.S. Department of Education, Office of Special Education and Rehabilitative Services. (2005). www.ed.gov/about/offices/list/osers/index.html

U.S. 94th Congress. (1975). Public Law 94.142.

Winnick, J. (2000). *Adapted physical education and sport*. Champaign, IL: Human Kinetics.

ADDITIONAL RESOURCES

In addition to the chapter references, the organizations listed here are good sources of information on adapted activities for the disabled.

American Association of Adapted Sports, www.aaasp.org

American Association of the Deaf-Blind, www.aadb.org

American Academy for Cerebral Palsy and Developmental Medicine, www.aacpdm.org

American Alliance for Health, Physical Education, Recreation, and Dance, www.aahperd.org

American College of Sports Medicine, www.acsm.org

American Congress of Rehabilitation Medicine, www.acrm.org

American Dance Therapy Association, www.adta.org

American Diabetes Association, www.diabetes.org

American Epilepsy Society, www.aesnet.org

American Foundation for the Blind, www.afb.org

American Heart Association, www.americanheart.org

American Medical Association, www.ama-assn.org

American Music Therapy Association, www.musictherapy.org

American Occupational Therapy Association, www.aota.org

American Physical Therapy Association, www.apta.org

American Psychiatric Association, www.psych.org

American Psychological Association, www.apa.org

An Association for Children and Adults with Learning Disabilities, www.acldonline.org

Association for Retarded Citizens, www.thearc.org

British Wheelchair Sports Federation, www.bwsf.org.uk

Council for Exceptional Children, www.cec.sped.org

Disabled Sports USA, www.dsusa.org

Epilepsy Foundation, www.epilepsyfoundation.org

Joseph P. Kennedy, Jr. Foundation, www.jpkf.org

March of Dimes Birth Defects Foundation, Division of Health Information and School Relations, www.modimes.org

Muscular Dystrophy Association, www.mdausa.org

National Association for Visually Handicapped, www.navh.org

National Association of the Deaf, www.nad.org

National Association of the Physically Handicapped, www.naph.net

National Hemophilia Foundation, www.hemophilia.org

National Mental Health Association, www.nmha.org

National Multiple Sclerosis Society, www.nmss.org

Special Olympics, www.specialolympics.org

Spina Bifida Association of America, www.sbaa.org

United Cerebral Palsy, www.ucp.org

United States Association of Blind Athletes, www.usaba.org

U.S. Department of Education, Office of Special Education and Rehabilitative Services, www.ed.gov/about/offices/list/osers/index.html

USA Deaf Sports Federation, www.usdeafsports.org

Wheelchair Sports Worldwide, www.wsw.org.uk

10

EXTRACURRICULAR PROGRAMS
INTRAMURALS AND
INTERSCHOLASTIC ATHLETICS

Outcomes

After reading and studying this chapter, you should be able to:

- Define:
  - *Coaching certification*
  - *Exclusionary varsity model*
  - *Interscholastic athletics*
  - *Intramurals*
  - *Pay-to-play*
  - *Role conflict*
  - *Sport*
  - *Sport specialization*
- Distinguish between an intramural and an interscholastic athletic program.
- Discuss the importance of sports in our society.
- Explain how sports may affect the moral and ethical behavior of students.
- Discuss current interscholastic athletic issues including coaching certification, the pay-to-play policy, and specialization.
- Justify the need for both an intramural program and an interscholastic athletic program.
- Discuss both the strengths and weaknesses of an interscholastic athletic program.
- Discuss the effect that role conflict can have on both the physical education curriculum and the interscholastic athletic program.
- Discuss the effect the exclusionary varsity model can have on students.

Sport is as old as civilization itself. Its prominence, especially in Western cultures, is signified by the ever-increasing number of participants and spectators, the increasing number of sports publications, and the number of radio and television programs geared solely towards sports. Siedentop (2004) notes that sports have an almost religious significance in American culture. During the seventh game of the World Series, throughout the NCAA Final Four, or on Super Bowl Sunday, the nation seems to pause momentarily as attention is focused on these sporting events.

Our American society places far more emphasis on its sports than, perhaps, any civilization since ancient Rome. In fact, sports have attained such a prominent position of influence in American society that propagation now involves nearly the total socialization of its many participants. Sports have become an element of American life so pervasive that virtually no individual is untouched by them directly or indirectly.

It is apparent that sports are far more than a recreational diversion in American life. They are a social institution consisting of numerous complex and varied activities, values, and relationships. As a social institution, sports function in the development and reinforcement of appropriate values, the regulation of acceptable behavior, and the attainment of desired goals with the chief focus on the quality of performance by the participants.

Sports have impressive quantitative and qualitative aspects that enable them to be a major force in the promotion of acceptable practices and attitudes—if presented appropriately. Some concern exists today that too many youth programs are imperfect copies of professional sports in which the business is to win. Evidence suggests that as school sports are conducted in a more serious and businesslike fashion, these positive aspects may exceed the grasp of too many participants.

In the days of the early Greeks, both Spartan and Athenian philosophers focused on the dangers inherent in overemphasis and specialization in athletics. The intellectuals tolerated sports chiefly because they (1) prepared male citizens to defend the land from external aggression, (2) contributed to the advancement of the physical condition, and (3) fostered unity as teams came together to compete in national athletic festivities. Although such major sporting and game activities were accepted by the Romans, the concern persisted that sports could get out of control and defeat the sportsmanship objective. As a critic of society, Cicero saw some real value in sport but warned that it also symbolized, in a general way, the moral degradation of Roman society.

A well-conceived school sports program allows students to not only become aware of their limits and concepts of self, but also to appreciate a quality performance, in terms of both lasting skills and the essence of fair play. In this respect, participation in appropriate extra-class sporting activities is sound educational practice. This part of the curriculum is a well-tested experience consistent with the broad purposes of education.

The benefits and satisfaction derived from sports participation during the upper elementary grades (4–6) and the junior and senior high school years often carry over into adult life. Skilled, unskilled, and exceptional students all need the opportunity to participate in extra-class sports, whether through intramural and/or interscholastic programs.

MORAL AND ETHICAL BEHAVIOR

Sports are popular both in and out of school. Sport sociologists conduct ongoing research because of the positive and negative implications arising from sports participation. Moreover, the popularity of sports and people's involvement with sports will likely continue to grow as the interest in sport science and health-related fitness continues to increase. However, the effect that sports participation has on a student's affective development is a concern.

For a number of years researchers have been seeking more productive ways to influence the moral and ethical behavior of youth through sports participation. Sport has long been considered one of the best venues in which to learn about moral and ethical behavior. However, criticism is mounting regarding the unethical abuses in sports. In both intramurals and interscholastic athletics, some coaches place an overriding emphasis on winning at any cost, which diminishes the experience for many players, parents, and spectators. Striving to win a sport contest while exhibiting morally responsible behavior has always been an appropriate educational goal. Beyond that, the real issue is teaching values in connection with meeting student needs.

When not unduly influenced by overenthusiastic parents and coaches, most young people indicate that they would rather play on a losing team than sit on the bench of a winning team. Thus, if critics identify intramural and/or interscholastic athletic programs as having too much emphasis on competition and winning, then educators must clarify that the sports are scheduled as an extension of educational goals. As such, these programs are designed to meet the same objectives as the instructional program, and are capable of doing so through appropriate use of both competitive and cooperative processes.

Competition between opponents occurs when rules are enforced and tactics are employed; cooperation among teammates occurs when their overall responsibilities mesh and they are able to coordinate and execute their tasks. Competition and coordination are enhanced when moral and ethical behavior are emphasized in physical education programs in schools. The school's concern for responsible behavior also should be extended to parents and others in the community. Highly structured leagues and overenthusiastic parents are not always helpful in promoting ethical behavior. Parents are more effective when they motivate students to participate in sports and recreation, and reinforce appropriate social behaviors in these settings.

BALANCE IN PROGRAMMING

For all students, boys and girls alike, to have equal opportunity to participate in a variety of sport-related activities, a careful balance must be maintained between offerings in intramurals and interscholastic athletics. Conflict between these two programs should not occur if the resources (budgets, time, facilities, equipment, and personnel) are apportioned impartially—that is, according to how well each meets the overall objectives of the total program. Ideally, intramurals and interscholastic athletics can contribute to and complement one another. An overall program that is well-balanced allows students to enjoy participation and develop a respect for sports and the potential they have for improving one's physical, mental, social, and emotional development.

If a proper balance between programs at both the junior and senior high school levels is not maintained, then program emphasis will center primarily on interscholastic athletics. This is a natural occurrence in localities where school and community spirit run high and people take pride in their sports teams. Unfortunately, this imbalance will cause one part of the overall program to overshadow the others. When you consider that 100 percent of the students are exposed to the instructional physical education program, while as few as 15 to 20 percent take part in interscholastic activities, this overemphasis on sports at the expense of intramural and instructional programs is particularly ill-considered. The remaining 80 to 85 percent need encouragement and a chance to participate in extra-class activities suited to their abilities and interests. The traditional model illustrating the relationships among the three programs is shown in figure 10.1.

As shown, the instructional program provides the base of the pyramid and reaches every student in the school. The second level of the pyramid is the intramural program, which provides opportunities for students to further their skill development. The top of the pyramid is the interscholastic athletic program and it reaches the fewest number of students, although these individuals will be more highly skilled.

Intramural experiences are capable of promoting lasting values. They are completely voluntary and allow participants the opportunity to work at achieving a common objective. Therefore, the inclusion of these experiences in the program is both educationally sound and important.

Figure 10.1
Model for Programming Physical Education

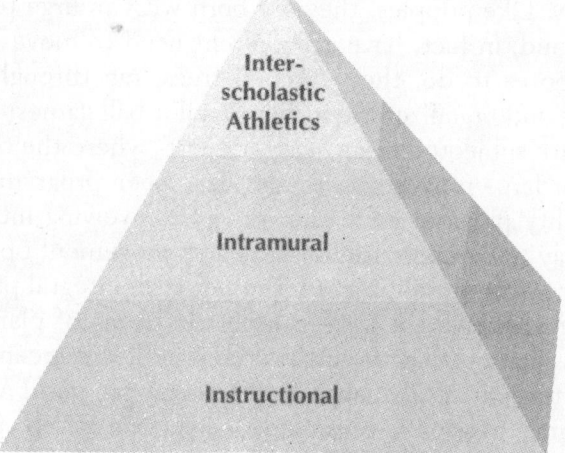

THE NEED FOR PLANNING

An extra-class program must be carefully planned in order to yield good results. This is especially true in large school systems where the cocurricular activities are extensive in both number and quality. Too often, intramurals and interscholastic athletics compete with each other for the student's free time, requiring such a concentration of effort that to participate fully in more than one or two experiences is nearly impossible. For example, a member of the school's band can hardly engage in varsity football, nor can a cross-country runner devote much attention to an after-school photography club.

Another factor affecting the programming of extra-class activities is the growing number of students committed to working after school hours. A lethargic economy and the desire among many students to enter the adult world as soon as possible have led to a high percentage of secondary school students preferring to work rather than take part in after-school activities. Despite the fact that many other attractive after-school activities exist in the upper elementary, junior, and senior high schools, it is possible to design extra-class physical education experiences that will reach students. The planning committee's task is to structure components of extra-class programs so that they are an extension of and supplemental to the developmental instruction program.

THE INTRAMURAL PROGRAM

It has been said that children are permanently tuned into physical activity. Like puppies, they are born with an urge to play. Children like to move and, in fact, have an inherent need to move. Decades ago, children had chores to do, they climbed trees, ran through the fields, skated on ponds, and organized their own sandlot ball games. Today's children, however, are subjected to an "easy society" where the once-natural opportunities for large muscle movement have been programmed out of their lives. A quality physical education program involving intramural experiences is one way to compensate for these lost movement opportunities.

The intramural program should be an integral part of the total curriculum and receive the same support in terms of planning, implementation, and evaluation that the instructional program receives. It should be a logical extension of a quality instructional program. While the instructional program's mission is to provide instruction for all students, the intramural program exists to provide students an opportunity to use and/or further refine the skills learned in the instructional program. Intramurals can be seen as a laboratory period for sports and other physical activities whose fundamentals have been taught in the regularly scheduled physical education program. The intramural program, which is voluntary in nature, should provide activity opportunities for all students, male and female, skilled and unskilled, big and small. Emphasis should be placed on skill application (play) with an instructional component relative to rules and the application of strategy, if necessary. A quality intramural program should enrich the expression of this natural, inherent urge to play. The unique purposes of the intramural program are to:

- Provide an opportunity for students to be physically active in the sport of choice at their level of competency
- Provide additional school time for skill development in activities taught in the instructional program
- Promote cooperation and socialization through participation in scheduled activities
- Promote sportsmanship, self-management, and self-control

This list reflects the ideal program. As Siedentop (2004) indicates, the ideal is seldom achieved. In fact, too often the intramural program is nonexistent due to insufficient time, facilities, personnel, and budget.

Providing a time for intramural activities is a frequent problem. Intramural activities may be scheduled before school, during the school day, after school, on weekends, and during vacation periods. Early morning programs are becoming increasingly popular as students opt to begin their day with activity. Intramural activities are often scheduled during the school day either over an extended lunch hour or during a regularly scheduled intramural period. Although used sparingly, weekend and vacation time scheduling is an option. This demands a strong commitment on the part of not only those directing and supervising the program, but also on the students. Perhaps the best time for scheduling intramurals is immediately after school; however, this is also the most convenient time for scheduling *all* other extracurricular activities, including interscholastic athletics and club activities. This compounds the problem not only in terms of competing for students but also in competing for facilities.

In order for the intramural program to be successful and to thrive, it needs strong leadership. This leader, commonly a physical educator in the school system, should be paid and given specified responsibilities. These responsibilities include the following:

- Prepare and administer the budget
- Chair the committees specifically responsible for the selection, organization, and implementation of activities
- Arrange schedules
- Evaluate the program on a regular basis
- Promote the program

Extra-class sports programs should be an extension of and supplemental to the developmental instruction program.

One duty that warrants further explanation is the topic of dealing with program awareness. Regardless of how well the activities are planned, if students do not participate then the intramural program will eventually fail. The following promotional techniques are designed to ensure public awareness:

- Provide announcements/schedules through as many media venues as possible, including television, radio, local and/or school newspapers, newsletters, school announcements, and bulletin boards.
- Include faculty by scheduling contests between and among students and teachers.
- Allow parents to attend scheduled events.
- Incorporate an awards system that gives recognition to all participants.

Additional guidelines to foster a successful program are listed in table 10.1.

Intramural sports can be presented in two different formats—organized and self-directed. The organized format is the most common. Typically students sign up to participate in a particular sport or physical activity and are subsequently assigned to teams. The teams are placed in leagues and then play in some tournament format with awards at the end. This format is most appropriate for team sports like basketball, soccer, and volleyball. The organized format is also appropriate for selected individual/dual activities such as badminton, golf, and tennis.

The self-directed format is a more informal approach. Students are still required to sign up for a specific activity. Following the sign-up stage, students are then provided with (1) a list of participants, (2) a required number of participants to compete against, and (3) a time frame within which to complete the number of games. It then becomes students' responsibility to contact opponents and play. This format is best used with individual/dual activities.

The selection of activities for the intramural program should be carefully planned. Recall that the intramural program, although extra-class, still needs to provide wholesome educational experiences. The intramural program should be an extension of the instructional program. Table 10.2 lists some of the more traditional activities that are appropriate for students in grades 4–12. Table 10.3 is a listing of intramural activities for middle school students from the Newport News public school system. This comprehensive program gives students an opportunity to participate in activities to:

- Acquire knowledge and refine skills required for future participation
- Improve self-esteem
- Stimulate an interest in selected physical activities
- Develop sportsmanship
- Enhance social interaction in a play environment

Table 10.1
Intramural Guidelines

- The program should supplement and not replace either the instructional or the inter-scholastic athletic program.
- The program should be broad enough in scope to interest and attract students.
- The activities selected should afford individual and group experiences for both boys and girls.
- The program should include corecreational as well as both individual and team activities for disabled students.
- The program should include both organized and self-directed activities.
- No student should be exempt from or denied the opportunity to participate in the program due to low academics or poor skill ability.
- Competition should be arranged between students of equal skill whenever possible.
- Individual and group instruction should be provided on students' requests.
- The program should be conducted according to a regular schedule.
- The program should make use of more than one available time period whenever possible (before, during, and after school) to better accommodate students.
- Officials should be well trained and qualified to maintain safety and promote fair play.
- Complaints and protests should be handled in an equitable and democratic manner.
- Tournaments that offer continuous participation (round robin and ladder) rather than limited participation (single elimination and double elimination) should be utilized.
- Intramural records should be kept not only for public relations purposes but also for making program evaluations.
- The program should have an awards system that recognizes all participants.
- The program should not be perceived as a proving ground for interscholastic athletic candidates.

Table 10.2
Seasonal Intramural Activities for Students in Grades 4–12

Fall	Winter	Spring
Archery	Badminton	Golf
Backpacking	Basketball	Horseshoes
Bicycling	Bowling	Miniature golf
Field hockey	Floor hockey	Shuffleboard
Flag football	Gymnastics	Softball
Flickerball	One-on-one basketball	Tennis
Jogging	Roller skating	Track and field
Soccer	Swimming	
Speedball	Table tennis	
	Volleyball	
	Wrestling	

Table 10.3
Newport News Yearly Intramural Schedule

Session 1 (September 15–November 13)	Session 2 (January 5–March 15)	Session 3 March 22–May 13)
Field hockey	Badminton	Croquet
Flag football	Basketball	Frisbee golf
Speedball	Floor hockey	Horseshoes
Soccer	Pickleball	Pickleball
Tennis	Volleyball	Softball
*Culminating activity: Punt, Pass, & Kick competition	*Culminating activity: Hot Shot contest	Tennis
		Track/Cross country
		*Culminating activity: Fun Fitness Day

All sessions are held after school with transportation provided. Ongoing activities include an aerobics club, cycling club, jogging club, and a walking club.

In addition to the more traditional activities, some argue that offering novel activities is one means of increasing participation. These less traditional offerings allow students to experience activities not included in the instructional program.

THE INTERSCHOLASTIC ATHLETIC PROGRAM

In many communities, the crowning achievement of the overall physical education curriculum is the opportunity provided to physically gifted students to compete in interscholastic athletics. This program undoubtedly provides both boys and girls a valuable educational experience. School personnel—teachers and administrators alike—who have constructed and soundly implemented a K–12 physical education curriculum should perceive the interscholastic athletic program as advancing the limits of an intramural program to a more competitive, more demanding, and more concentrated experience.

Siedentop (2004) indicates that the large, diversified interscholastic athletic program is unique to American junior high schools and senior high schools. Most secondary schools have teams at the freshman, junior varsity, and varsity levels in a variety of sports. The importance that interscholastic athletics play in our culture and the ultimate effect they have on our youth regardless of whether they compete is undeniable.

During the 2003–2004 school year, nearly 7 million students, 4 million males and 2.9 million females, participated in more than 40 different high school sports (see table 10.4). This number accounts for 53.3 percent of those enrolled in high school. The National Federation of State High

School Associations (2004) reported that participation in high school athletics increased for the sixth consecutive year. The federation further indicated that this increase is partly due to a record number of female participants. The number of male and female participants for the ten most popular sports is shown in tables 10.5 and 10.6, respectively.

The interscholastic program provides an opportunity for physically gifted students to compete at the highest level of ability against boys and girls from other schools. This competition is often under the auspices of the school and an interscholastic athletic conference, if not the state. Well-organized and properly controlled competition between schools can make a valuable contribution to the overall goals of education but more specifically to the goals of physical education. The more commonly accepted purposes of an interscholastic athletic program are to:

- Contribute to the maximum development of health-related fitness
- Contribute to the maximum development of motor fitness
- Contribute to self-discipline and self-control

Table 10.4
Interscholastic High School Sports

Archery	Equestrian	Lacrosse	Spirit squad
Badminton	Fencing	Riflery	Swimming
Baseball	Field hockey	Rodeo	Team tennis
Basketball	Flag football	Sailing	Tennis
Bowling	Football	Skiing (alpine)	Track & field (outdoor)
Canoeing	Golf	Skiing (cross country)	Track and field (indoor)
Crew	Gymnastics	Snow boarding	Volleyball
Cross country	Ice hockey	Soccer	Water polo
Dance team	Judo	Soft tennis	Weight lifting
Diving	Kayaking	Softball (fast and slow pitch)	Wrestling

Source: National Federation of State High School Associations, http://www.nfhs.org/ScriptContent/Index.cfm

Table 10.5
Most Popular Male Sports

Sport	Participants
1. Football	1,032,683
2. Basketball	544,811
3. Track and field (outdoor)	504,801
4. Baseball	457,146
5. Soccer	349,785
6. Wrestling	238,700
7. Cross country	196,428
8. Golf	163,341
9. Tennis	152,938
10. Swimming	96,562

Table 10.6
Most Popular Female Sports

Sport	Participants
1. Basketball	457,986
2. Track and field	418,322
3. Volleyball	396,322
4. Softball (fast pitch)	362,468
5. Soccer	309,032
6. Tennis	167,758
7. Cross country	166,287
8. Swimming and diving	144,565
9. Spirit squad	89,443
10. Golf	63,173

A well-organized and implemented interscholastic athletic program can make a valuable contribution to the overall goals of education and, in particular, physical education.

- Provide an opportunity to develop leadership skills
- Provide an opportunity to compete with others with similar abilities in sports
- Contribute to school loyalty

Unlike the instructional and intramural programs, interscholastic athletics are characterized by a higher degree of organization, an increased number of spectators, considerable publicity, and too often, commercialism. As a result, interscholastic athletic programs sometimes require close scrutiny. Two specific concerns arise: first, the interscholastic athletic program should be included in the curriculum only after both the instructional and intramural programs are firmly established. Second, due to the emotional atmosphere surrounding a competitive event and the desire to win by players and coaches alike, athletics are more subject to manipulation than either the instructional or intramural programs. This overemphasis on winning and the resultant pressure on coaches may lead to player exploitation and other abuses that can threaten the entire physical education curriculum. The interscholastic athletic program should not be viewed as either community entertainment or a revenue source for the school. It is nothing more than an opportunity for students to pit their sports skills and knowledge against students with equal or greater ability.

The values and benefits derived from the interscholastic athletic program do not come automatically from participation. These experiences must be well planned under the leadership of knowledgeable, dedicated school personnel and be well established on sound policies whether they are developed by the school, the athletic conference, or the National Federation. Even in the best of situations, problems arise. Jewett, Bain, and Ennis (1995) report that from a historical perspective, physical education and athletics have been closely related. From a practical viewpoint, both programs have shared facilities, if not equipment. In addition, it is not uncommon to find that both programs are conducted by the same individuals, serving as both physical educators and coaches. One or both of these situations has led to the following problems:

- The athletic program abandoning educational goals for an emphasis on winning
- The physical education program being used to identify and/or develop potential athletes
- Athletic participation serving as a substitute for the physical education requirement
- Diminished administrative support for physical education as athletic visibility increases
- Role conflict as the physical educator attempts to be both teacher and coach
- The development of the exclusionary varsity model (Siedentop, 2004)

Two of these problems—role conflict and the exclusionary varsity model—will be discussed in detail in the following paragraphs.

Many individuals choosing to coach in today's schools are hired as teachers. Therefore, they are filling two similar but distinctly different roles and the result is role conflict, a condition that arises when incompatible expectations for different roles exist (Sage, 1987). Figone (1994) states that role conflict is the result of a teacher-coach attempting to fulfill the expressed expectations of both roles. Too often the individual either (1) falls short of the expectations of both roles; or, in most instances, (2) devotes time and energy to one role thereby neglecting the other. Role conflict flourishes because the rewards and expectations for coaching are so much greater than for teaching. Coaches are in the public eye. Even though both the teacher and the coach come under close scrutiny by administrators and parents, too often the teacher's accomplishments go unnoticed while the coach is admired and praised. Therefore, it is easy to understand why the teacher-coach devotes more time to the coaching role.

The exclusionary varsity model refers to the reality that fewer students are able to participate in varsity athletics than in other sports programs. Sports participation for elementary school children is generally community based through YMCA, YWCA, and similar community-wide recreation leagues. As Siedentop (2004) indicates, these programs are, for the most part, widely available and open to children of all skill levels. Sports participation for secondary school students is programmed primarily by the schools. In addition, this participation is geared toward varsity participation, which is exclusionary in nature—it acts to identify and cater to the highly skilled athlete. The secondary school students excluded from varsity-level athletics have few opportunities to continue their sports participation, at least until they are old enough to take part in a community-based adult program. Unless an adolescent is highly skilled, the result is little opportunity to further develop those skills the individual had an interest in developing. Further delineation of both the strengths and weaknesses of an interscholastic athletic program is provided in table 10.7.

The selection of activities (sports) for the interscholastic athletic program may come from the list of sports presently sponsored in secondary schools throughout the country (see table 10.4). This list is rather extensive because regional and local needs, interests, and facilities may vary some-

---

Table 10.7
Strengths and Weaknesses of Interscholastic Athletics

Strengths

Athletics:

- Act as a unifying agent leading to group loyalty
- Provide self-testing opportunities
- Provide opportunities for students to interact with students from other schools
- Allow students, both as spectators and participants, to learn how to function in a sports-conscious society
- Provide a place for physically gifted students to meet their need for competition
- Provide an opportunity for the development of sportsmanship, self-control, and self-responsibility

Weaknesses

Athletics:

- Emphasize the physical development of a few skilled performers while neglecting the ordinary student
- Can disrupt the regular school routine with contests during school hours
- Can encourage gang-like behavior, resulting from interschool rivalries
- Can place undue pressure on both coaches and athletes due to a win-at-all-cost attitude
- When overemphasized, may adversely impact the instructional and/or the intramural program
- May lead to commercialism
- May adversely affect the academic growth of students because of the time commitment they require

Source: Adapted from Annarino, Cowell, and Hazelton, 1986.

what according to climate, geography, and traditions. The total number of interscholastic sports offered in any one school depends primarily on budget, student interest, and the availability of competition. For example, if the budget allows for and student interest warrants the inclusion of soccer in the interscholastic athletic program, but there are no opponents within a 100-mile radius, it would not be logical to include soccer in the program.

The appropriateness of athletics for junior high school students has been discussed for years. The effects of both collision sports (football, ice hockey, and lacrosse) and contact sports (basketball, soccer, and wrestling) and the undesirable pressures from intense athletic competition have been debated. The stand taken by many sport leaders is that highly organized interscholastic athletic programs are questionable for junior high school students. Many contend that the main concentration at the junior high school level should be on a broad intramural program. These young people are in a transitional stage between elementary school and high school. It is at this time that students are gaining an increased interest in sports, but their physical and social immaturity make it unwise for them to participate in an interscholastic athletic program. This contention, however, has not significantly lessened the number of junior high school athletic programs. Neither has the fact that junior high school programs face a set of problems not found in a good intramural program, including unqualified coaches, specialization at ages that are too young, and developmentally inappropriate sports. If interscholastic sports are intended to be a part of the junior high school program, it is essential that they be based on sound principles. Bucher and Koenig (1983) propose the following four guidelines:

1. The primary objective should be healthful participation.
2. A certified teacher with knowledge of both the participant and the sport should provide leadership.
3. Students should be encouraged to participate in many sports rather than specialize in just one.
4. Selected sports should be modified pertaining to the developmental age of the participants.

These principles apply to athletics at any educational level, but especially at the junior high school level.

In the past two decades a variety of trends have impacted high school athletics. Four such trends are highlighted in figure 10.2 and will be discussed next.

Coaching Certification

The United States is the world's only major sporting nation without a certification program for its coaches. A means of ensuring that student athletes are under the leadership of qualified individuals is necessary at the state level if not also at the national level. As previously stated, nearly 7

Figure 10.2
Trends in Interscholastic Sports

million students in the United States participated in interscholastic sports in 2003–2004, playing on more than 200,000 teams. The number of coaches required to staff this many teams is substantial. The majority of these students are coached by individuals who are certified just to teach. However, this has changed dramatically since the number of certified teachers who want to coach is fairly low. The result is what the American Sport Education Program (1996) has termed the *accidental occupation*, meaning that coaching in U.S. schools is an occupation that many have drifted into rather than something they have prepared for. In fact, it is estimated that more than two-thirds of interscholastic coaches have received little formal coaching education, if any at all.

There are no national certification requirements for someone wanting to coach at the secondary school level. Although requirements for coaches do exist in some states, they vary from state to state and are often related to teacher certification, not to coaching preparation. Siedentop (2004) indicates that some states have no standards governing the hiring of coaches, while other states leave this to either the state board of education or the state activities association. With the expansion of sports at all levels and the continued growth of interscholastic programs for girls and women, a shortage of qualified coaches is present. This shortage has resulted in a steady increase in the number of unqualified personnel coaching today's youth.

In 1992 the National Association for Sport and Physical Education (NASPE) appointed a special task force to consider ways of improving the quality of coaching in the United States. The result was the development and publication of *National Standards for Athletic Coaches* (NASPE, 1995). The standards are intended to provide direction for administrators, coaches, athletes, and parents regarding the skills and knowledge that coaches should possess. These standards (1) are a compilation of the knowledge, skills, and values associated with effective coaching; (2) reflect the basic competencies expected of coaches at any level; and (3) are to be used to ensure the enjoyment, safety, and skill development of today's athletes. These 37 standards are grouped into the following categories:

- Injuries: prevention, care, and management
- Risk management
- Growth, development, and learning
- Training, conditioning, and nutrition
- Social/psychological aspects of coaching
- Skills, tactics, and strategies
- Teaching and administrative aspects
- Professional preparation and development

The National Council for Accreditation for Coaching Education (NCACE) held its inaugural meeting in 2000 and is partnered with NASPE. NCACE's mission is to support qualified coaches through programs that provide quality coaching education. Its primary function is to review coaching education and certification programs that seek accreditation. Support for this effort has been widespread, including the National Alliance for Youth Sports, the National Association for Girls and Women in Sports, the National Association for Sport and Physical Education, the National Federation of State High School Associations, the United States Sports Academy, the Women's Sport Foundation, and the Youth Sports Institute.

Another attempt to address the issue of coaching certification has been the development and marketing of coaching education programs. Three such programs include:

1. American Sport Education Program (ASEP)
   http://www.asep.com
2. National Youth Sports Coaches Association (NYSCA)
   http://www.nays.org/
3. Program for Athletic Coaches Education (PACE)
   http://www.mhsaa.com/administration/pace.html

The American Sport Education Program (ASEP) began in 1976 as the American Coaching Effectiveness Program. Although coaching education is ASEP's priority, it also provides education programs for officials, sport administrators, parents, and athletes. ASEP has a five-level program shown in table 10.8. Levels 1 and 2 concern volunteer coaches, while levels 3, 4, and 5 comprise the professional education program for high school, college, Olympic, and club sport coaches.

The National Youth Sports Coaches Association (NYSCA) is a nonprofit membership organization designed to educate coaches about their importance in the physical, psychological, and emotional development of youths. It has trained more than 1.8 million coaches since its inception in 1981. Its primary objective is to sensitize volunteer coaches to their responsibilities and hold them accountable to the NYSCA Coaches' Code of Ethics (see table 10.9).

NYSCA has an Initial Level Membership (ILM) and Gold Level Certification (GLC). ILM requires coaches to complete an interactive video training clinic, pass an exam, and pledge to uphold the Coaches' Code of Ethics. The clinic includes course work in coaching philosophy, effective practices, sport specific fundamentals, and injury prevention and treatment. To achieve GLC, coaches must complete an online course dealing with eight content areas:

- Philosophy and ethics
- Sport safety and injury prevention
- Physical preparation and conditioning
- Growth and development
- Teaching and communication
- Organization and administration
- Skills and tactics
- Evaluation

Table 10.8
ASEP Coaching Programs

Level 1—Basic Education	A 3–5 hour course with content on coaching philosophy and principles.
Level 2—Beyond the Basics	A 6-hour online advanced course with content on coaching methodology.
Level 3—Bronze Certification	Granted to those who complete courses in coaching principles, first aid, and sport specific theory.
Level 4—Silver Certification	For those who have bronze certification and want further professional development. Granted to those who complete courses in sport physiology, sport psychology, sport mechanics, and sport skill instruction.
Level 5—Gold Certification	For those who have silver certification and want further professional development. Granted to those who complete course work in risk management and sport issues and advanced course work in sport sciences and skill instruction

Table 10.9
NYSCA Coaches' Code of Ethics

- I hereby pledge to live up to my certification as a NYSCA Coach by following the NYSCA Coaches' Code of Ethics.
- I will place the emotional and physical well-being of my players ahead of a personal desire to win.
- I will treat each player as an individual, remembering the large range of emotional and physical development for the same age group.
- I will do my best to provide a safe playing situation for my players.
- I will promise to review and practice the basic first aid principles needed to treat injuries of my players.
- I will do my best to organize practices that are fun and challenging for all my players.
- I will lead by example in demonstrating fair play and sportsmanship to all my players.
- I will provide a sports environment for my team that is free of drugs, tobacco, and alcohol, and I will refrain from their use at all youth sports events.
- I will be knowledgeable in the rules of each sport that I coach, and I will teach these rules to my players.
- I will use those coaching techniques appropriate for each of the skills that I teach.
- I will remember that I am a youth sports coach; the game is for children, not adults.

Source: National Youth Sports Coaches Association, http://www.nays.org/

The Program for Athletic Coaches Education (PACE) emanated from the establishment of the Michigan Institute for the Study of Youth Sports by the state legislature in 1978. The legislature proposed three mandates:

1. To conduct research dealing with the effects of sports on youth
2. To provide educational materials to those involved with youth sports
3. To provide educational programs for coaches, administrators, officials, and parents

PACE has a two-level program. Each level requires individuals to attend a one-day clinic and pass an exam with a score of 80 percent or higher. Level 1 content includes coaching guidelines and legal responsibilities as well as prevention, care, and rehabilitation of sport injuries. Level 2 content includes motivation, conditioning, and evaluation.

Few people would argue that a need exists for our sport-minded youth to learn in a wholesome educational environment. Furthermore, perhaps the most significant component of this environment is the coach. The development and implementation of the above-mentioned programs is an attempt to provide better qualified coaches for today's youth.

Pay-to-Play

Even in the best economic times, school districts across the United States find it difficult to provide a quality education with funds available.

School districts search to find creative ways of generating revenue to prevent programs from being reduced or eliminated altogether. One approach is to require a user's fee for extracurricular activities. Students who elect to participate in these activities, of which athletics is one, must pay for this experience, with fees ranging from $25 to $200 per sport. This pay-to-play policy is illegal in some states because it is seen as discriminatory to those students who lack the resources to participate.

A solid argument against user's fees is made by the National Federation of State High School Athletic Associations. The organization states that interscholastic athletics are included in the school's overall program because of their educational value. Therefore, if they are seen as such, athletics should be funded and made available to all students. Yet, because of the rising costs of administering a comprehensive athletic program, compounded with recurring budget cuts, user's fees may provide the only viable means of new revenue to help fund interscholastic athletic programs.

Specialization

The days of an athlete participating in three or four sports are all but over. With coaches demanding more from their athletes and athletes striving for a competitive edge, there is pressure on top athletes to specialize in one sport. Specialization involves athletes practicing, training for, and competing in their specialty throughout the year (Hill & Hansen, 1988). Specialization at the high school level has occurred as a result of two developments (Siedentop, 2004): (1) the national acceptance of off-season training, and (2) the increased availability of athletic scholarships at the college/university level. Watts (2002) states that proponents support specialization because it enhances athletic performance for both athletes and the teams on which they play, increases the probability of winning, and satisfies the athlete's desire to achieve excellence.

Sport as Entertainment

Professional-level sports provide entertainment for millions of Americans. This view has also reached the collegiate level as millions of dollars are spent annually on the marketing and televising of college athletics. On a lesser scale, this attitude has now reached the high school level and, as a result, interscholastic athletics are becoming a business. This may prompt administrators to look at sports as a means of producing revenue for the school rather than as an opportunity to provide a more well-rounded education to students. As a result, the interscholastic program may be adversely affected.

SUMMARY

1. Sports in contemporary America have achieved an unparalleled prominence.

2. Interscholastic sports involvement in the U.S. is unmatched by any other country, with nearly 7 million participants during the 2003–2004 school year.

3. The intramural program, often viewed as an extension of the instructional program, must be accepted as integral to the physical education curriculum.

4. Intramurals can be presented in an organized and/or self-directed format.

5. Sports, whether scheduled through an intramural or an interscholastic athletic program, are educational—that is, they are designed to achieve predetermined objectives.

6. A balance needs to be maintained between intramural and interscholastic athletic programs.

7. For educational benefits to be derived from participation in interscholastic athletics, a need for both leadership and planning exists.

8. Various trends have impacted interscholastic athletics, including coaching certification, the pay-for-play policy, specialization, and sport as entertainment.

QUESTIONS AND LEARNING ACTIVITIES

1. Why do athletes seem to belong to more clubs, have wider interests, and be more extroverted and better adjusted socially then nonathletes? Does participation in athletics develop these qualities or do students with these qualities tend to participate in athletics?

2. Should a school sponsor a wide variety of intramural and interscholastic athletic activities if the community has a fairly extensive program (YMCA, YWCA, etc.)? Explain your answer.

3. It has been said that it is no longer a desirable practice to divide sports into established seasons. What may be the reasons for this viewpoint?

4. Interview two or three directors of physical education about broadening the girls' intramural and athletic programs. What problems have arisen as a result of compliance with Title IX regulations? Interview two or three women coaches about the same problem. How do their answers compare with those of the directors of physical education?

5. Indicate why it is important that coaches, players, students, and the community understand the school's philosophy of interscholastic athletics.

6. Interview three or more high school athletic directors regarding what they view as the key issues/concerns in today's interscholastic athletics.

7. How can interscholastic athletics affect the development of ethical behavior? In answering this question, list both the positive and the negative implications arising from sports in today's society.

8. Provide support for the placement of the three components of the physical education curriculum as illustrated in figure 10.1.

9. If you were to develop a coaching certification program, what criteria would you require coaches to meet?

REFERENCES

Bucher, C., & Koenig, C. (1983). *Methods and materials for secondary school physical education.* St. Louis, MO: Mosby.

Figone, A. (1994). Teacher-coach role conflict: Its impact on students and student-athletes. *The Physical Educator, 51*(1), 29–34.

Hill, G., & Hansen, G. (1988). Specialization in high school sports: The pros and cons. *Journal of Physical Education, Recreation, and Dance, 59*(5), 76–79.

Jewett, A., Bain, L., & Ennis, C. D. (1995). *The curriculum process in physical education.* Dubuque, IA: Brown & Benchmark.

National Association for Sport and Physical Education. (1995). *National standards for athletic coaches.* Washington, DC: AAHPERD.

National Federation of State High School Associations. (2004). Participation Survey, 2003–2004. http://www.nfhs.org/ScriptContent/Index.cfm

Sage, G. (1987). The social world of high school athletic coaches: Multiple role demands and their consequences. *Sociology of Sports Journal, 4,* 213–228.

Siedentop, D. (2004). *Introduction to physical education, fitness, and sport.* Champaign, IL: Human Kinetics.

Watts, J. (2002). Perspectives on sport specialization. *Journal of Physical Education, Recreation, and Dance, 73*(8), 32–37, 50.

11

CURRICULUM EVALUATION

Outcomes

After reading and studying this chapter, you should be able to:

- Define:

Affective domain	*Process assessment*
Cognitive domain	*Product assessment*
Evaluation	*Program evaluation*
Formal knowledge assessment	*Psychomotor domain*
Informal knowledge assessment	*Student evaluation*
Measurement	*Teacher evaluation*
Physical domain	*Test*

- Discuss the principles for guiding your approach to evaluation.
- Describe evaluation procedures that can be used to assess program effectiveness.
- Describe evaluation procedures that can be used to assess teaching effectiveness.
- Describe evaluation procedures that can be used to assess students in each of the four domains.
- Identify the four purposes for assessment of students in the adapted physical education program.
- Discuss why ongoing teacher evaluation is important.

A teacher's primary function is to bring about desirable behavioral changes in students. These behavioral changes should be based on stated educational objectives. The search for indications that these educational objectives have been achieved is the essential concept of evaluation and the process whereby teachers demonstrate accountability to parents and school board members.

As educators struggle to update our educational system—which is a never-ending process—they are becoming more and more accountable for their expenditures of both human and financial resources. No one can oppose accountability. However, it is entirely possible that the methods chosen to determine this accountability may impede or even destroy the goals of education. There is a need to think carefully about the *means* chosen for assessment in today's ever-changing world. As assessment techniques become more refined, evidence of student shortcomings should decrease, such as graduating from high school with the inability to complete an employment application. Fewer students should leave physical education programs with (1) the inability to perform lifetime sport skills, (2) an inadequate level of fitness, and (3) a lack of appreciation for the role that physical activity plays in the maintenance of a healthful lifestyle.

Today's physical education teachers are not always fully aware of shortcomings in their work. They do not struggle with cost-benefit programs, nor do they measure their successes by health care costs over a person's lifetime. The teaching profession does not have a *tremble factor*, a concept that was employed in ancient Rome. When the scaffolding of a building was removed from a completed arch, the Roman engineer responsible for its construction stood beneath it. If the arch came crashing down, then the engineer was the first to know. As a result, the concern for the quality of the arch was intensely personal, which may explain why so many Roman arches have survived for more than 2,000 years. If today's teachers had to stand under the educational *arch* set forth for the students, how many would survive the tremble factor?

DEFINITIONS

Whenever evaluation is discussed, it is difficult not to mention both tests and measurement. These three terms—evaluation, tests, and measurement—are often used interchangeably. They do not mean the same thing but are closely related. The following definitions are given to form a basis for the content that follows.

- *Measurement*. Simply, this is the process of collecting data.
- *Evaluation*. This is a decision-making process involving the collection of data and the determination of the worth of this data.

- *Test*. This is a form of assessment to measure the acquisition and retention of knowledge or ability in some mental or physical endeavor.

The relationship between these three terms can be best explained as the use of an instrument (test) for the collection of data (measurement) that allows one to make a more appropriate decision (evaluation).

THE INTENT OF MEASUREMENT AND EVALUATION

Measurement in education is related to preconceived aims and objectives. It is a process of making comparisons and relating them to one's needs in an effort to find what an individual has achieved.

To evaluate, as stated in the Merriam-Webster dictionary, is to appraise carefully, and to appraise is to establish a value. This value is determined by relative worth, excellence, or importance. The process of evaluating, therefore, should be considered along with measurement. Measurement answers the questions of how much, how many, and how often and is concerned with quantities and qualities in evidence. Evaluation goes beyond the mechanics of testing and measuring to render judgment in light of preconceived aims and objectives. Evaluation answers the question of whether a particular experience has value. It is a continuous process and should be fully integrated with the teaching-learning process. In its broadest sense, it concerns the advancement of the total school curriculum as well as the teachers and students the school serves.

The current emphasis is not placed too heavily on tests themselves but rather on the application of these tests for solving problems in physical education. In the end, therefore, measurement devices and evaluation techniques become an administrative means to aid teachers in helping their students. As a result, the most common purpose of measurement for evaluation in physical education is to determine the status, progress, and/or achievement of the individual student. Measurement also provides other valuable services; for example, it can be used to:

- Classify students
- Determine student status for grading
- Aid in the diagnosis of student weaknesses in fitness, skill development, and so on
- Predict student success
- Motivate students
- Determine program effectiveness
- Determine teacher effectiveness
- Contribute to research
- Improve public relations with students, colleagues, administrators, parents, and the community

EVALUATION GUIDELINES

Dunham (1994) explains that one's general approach to the instructional process and to evaluation within that process is dependent on one's philosophical perspective. This philosophy, which encompasses the values held concerning selected aspects of life, governs a person's actions. In physical education, an instructor's philosophy dictates not only the curriculum content selected but also the instructional procedures employed and the evaluation techniques utilized.

To ensure a more effective physical education curriculum, evaluation must be guided by sound principles. Dunham (1994) provides a comprehensive list of guidelines for physical education evaluation. Three important principles from this list indicate that evaluation should be:

- Accepted as an integral part of the teaching process
- Employed to assist students in achieving terminal competencies (psychomotor, cognitive, and affective)
- Based on the status of the individual student

AN OVERVIEW OF EVALUATION

The financial burden of supporting today's schools rests primarily on taxpayers. During inflationary times, schools are called on to justify the worth of their educational programs. To do this, both educators and administrators must continually evaluate all aspects of the program in order to (1) highlight areas that are performing well and (2) identify and strengthen areas that are not.

There is widespread concern at present for adopting higher standards in both public and private institutions. This has prompted new approaches to evaluation. Nationwide, schoolchildren are repeating grades and classes as school districts stiffen promotion and graduating standards. The notion today is to appraise progress carefully in all subject matter areas and work harder with those students needing special attention.

Current trends in physical education have brought about (1) increased use of fitness tests and (2) further study of the factors contributing to lifetime, health-related fitness. Motor development is being appraised through appropriate measures of general motor ability. Sport-specific skills are being measured. Written tests for the assessment of knowledge in physical education are now common. The quality of extra-class programs, intramurals, and interscholastic athletics is being measured in terms of personal student goals.

Total health behavior, academic achievement, self-concept, and personal happiness are being related more to physical performance and such specific factors as body mechanics, strength, muscular endurance, and aerobic endurance. Also significant is the fact that a majority of state departments of education have now suggested a course of study or curriculum guide with achievement standards or outcomes for sport skills, general motor ability, social behavior, and fitness. Those who employ such guides and courses of study are expected to compare their students with these standards in order to obtain some indication of student progress and future needs.

Some state departments of education have published competencies that outline the expectations of physical education programs. One such example is the Pennsylvania Department of Education's *Academic Standards for Health, Safety, and Physical Education* (see table 6.1), which lists competencies for grades 3, 6, 9, and 12. These competencies are minimal, but delineate what students should know and be able to do across these four grade levels.

In 2001 the North Carolina Department of Public Instruction revised its *Healthful Living Standard Course of Study and Grade Level Competencies* for health and physical education. The 12 competency goals are shown in table 11.1. The number of specific behavioral competencies within each goal varies from three to seven. Since goals 7–12 are specific to physical education, two competencies within each of these goals are provided.

Forethought is essential to the evaluation process. Planning and organization of time and resources are necessary to prevent a haphazard approach to the measurement of program effectiveness. It is a good practice to measure student health-related fitness twice a year—once at the beginning of the year for diagnostic assessment and again at the end of the year to determine student progress and both teacher and program effectiveness. Disabled students may need to have their progress checked more often.

Tests of skills and knowledge generally should be given at the end of each unit of instruction or at the end of each six- or nine-week grading session. It is recommended that students, especially in the middle school, junior high, and high school grades, be measured at least once in the affective domain—the domain dealing with the development of a sound understanding of the nature and value of the physical education experience. Ideally, the testing procedures should be as carefully planned and carried out as the curriculum itself. It is recommended that the testing procedures be shared with both students and their parents.

A complete evaluation of the curriculum involves an assessment of the student, the teacher, and the program. A discussion of each assessment element follows.

Table 11.1
North Carolina Healthful Living Curriculum, High School

Competency Goal

1. The learner will direct personal health behaviors in accordance with own health status and susceptibility to major health risks.

2. The learner will apply the skills of stress management to the prevention of serious health risks for self and others.

3. The learner will interpret health risks for self and others and corresponding protection measures.

4. The learner will apply relationship skills to the promotion of health and the prevention of risk.

5. The learner will apply behavior management skills to nutrition-related health concerns.

6. The learner will choose not to participate in substance use.

7. The learner will achieve and maintain an acceptable level of health-related fitness.

    7.03 Complete a health-related fitness test and assess personal level of physical fitness, including monitoring of the heart.

    7.05 Design and implement a personal activity program that demonstrates the relationship between physical activity, nutrition, and weight management.

8. The learner will exhibit regular physical activity.

    8.01 Identify resources in the community that can be accessed to maintain regular physical activity.

    8.05 Participate regularly in health-enhancing and personally rewarding physical activity outside the physical education class setting.

9. The learner will demonstrate an understanding and respect for differences among people in physical activity settings.

    9.01 Execute respect for individual differences in physical activity settings.

    9.02 Synthesize and evaluate knowledge about the role of physical activity in a diverse society.

10. The learner will demonstrate responsible personal and social behavior in physical activity settings.

    10.02 Set personal goals for the development of skills, knowledge, and social responsibility, and work independently toward those goals.

    10.03 Practice acceptable sportsmanship and fair play behaviors in physical activity settings.

11. The learner will participate successfully in a variety of movement forms and gain competence towards lifetime physical activities.

    11.01 Participate at a competent level in small-sided games in at least one team sport.

    11.02 Participate at a competent level in small-sided games in at least one individual or dual sport.

12. The learner will demonstrate a competent level of physical activity, sport, and fitness literacy.

    12.02 Demonstrate competence in basic offensive and defensive strategies or tactics in team, individual, and dual activities.

    12.03 Apply rules, biomechanical or movement principles, problem solving, and fitness concepts to game and movement settings.

Source: Adapted from *Healthful Living Standard Course of Study and Grade Level Competencies*, North Carolina Dept. of Public Instruction, www.dpi.state.nc.us/

STUDENT EVALUATION

The evaluation of student performance and progress is becoming an increasingly important and integral part of all physical education programs and has therefore become an essential responsibility of all teachers. In addition to measuring students to determine a grade, evaluation provides the data necessary for determining overall teacher and program effectiveness. In order to best accomplish this task, the student needs to be evaluated in four domains (see figure 11.1).

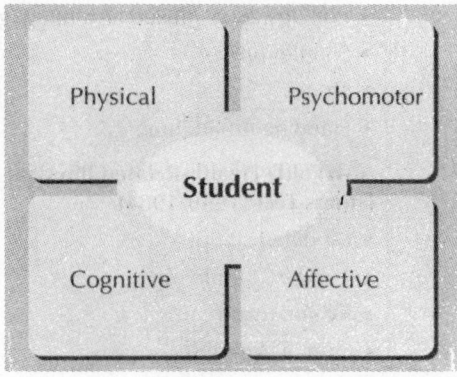

Figure 11.1
Domains for Student Evaluation

The Physical Domain

Assessment in this domain deals with growth and development and with the specific measures of height, weight, posture, and health-related fitness. It is highly recommended that the medical status of each student, as determined by a physician, be on file. This information may have diagnostic implications for the selection and/or organization of activities in the curriculum. The medical report should contain data relating to height, weight, and posture. The one area oftentimes not included in a medical report is fitness.

A student's level of fitness is strongly related to his or her success in fundamental locomotor skills and the basic game skills. Determining this level may result in significant adjustments to the curriculum. The growing popularity of aerobic (cardiorespiratory) activities and the research linking aerobic endurance to physiological well-being has sparked renewed interest in fitness assessment. As a result, a variety of fitness tests have been developed by physical educators, state departments of education, and colleges and universities (see table 11.2). The number of tests designed to measure so many different components of fitness make it is easy for the practitioner to become confused, but they nevertheless provide important feedback.

No group has done more work to develop a practical fitness test than AAHPERD. Its first attempt at developing pertinent fitness material was the Youth Fitness Test published in 1957, with revisions in 1965 and 1976. In 1980 AAHPERD developed the Health-Related Physical Fitness Test. Whereas the emphasis in the original test was placed on sports and motor skill performance, the new focus was on the components of fitness associ-

Table 11.2
Fitness Tests

AAHPERD Youth Fitness Test (1976)

- 50-yard dash
- 600-yard run
- Pull ups (boys)/flexed arm hang (girls)
- Shuttle run
- Sit ups
- Standing broad jump

AAHPERD Health-Related Physical Fitness Test (1980/1984)

- Modified sit ups
- 1-mile/1.5-mile run
- Sit and reach
- Sum of skinfolds

AAHPERD Physical Best (1988)

- Modified sit ups
- 1-walk/run
- Pull ups
- Sit and reach
- Sum of skinfolds

AAU Physical Fitness Test (1994)

- Distance run
- Pull ups
- Sit and reach
- Sit ups

ACSM Fitness Test (1998)

- Body mass index
- Push ups
- Rockport 1-mile walk
- Sit and reach
- Waist-to-hip ratio

Alabama Youth Fitness Test (1990)

- Curl ups
- 1-mile walk/run
- Pull ups
- Shuttle run
- V-sits

Chrysler Fund-Amateur Union Physical Fitness Test (1991)

Required

- Distance run
- Modified sit ups
- Pull ups (boys)/flexed arm hang (girls)

Optional

- Isometric push up (boys)/modified push ups (girls)
- Phantom chair
- Shuttle run
- Sprint 50/100 yards
- Standing broad jump

Fit Youth Today (1986)

- 20-minute steady state jog
- Curl ups
- Sit and reach
- Sum of skinfolds

President's Challenge (2000)

- Curl ups
- 1-mile walk/run
- Pull ups
- Shuttle run
- V-sit reach

Prudential FITNESSGRAM (1999)

- Back-saver sit and reach or shoulder stretch
- Curl up
- 1-mile walk/run or pacer
- Push ups, pull ups, modified pull ups, or flexed arm hang
- Sum of skinfolds or body mass index
- Trunk lift

YMCA Physical Fitness Test (1989)

- Bench press
- Bryde ergometer ride of 3-minute step test
- Sit and reach
- Sit ups
- Sum of skinfolds

ated with the prevention of disease and the promotion of functional health. These components included:

- Cardiorespiratory endurance
- Body composition
- Abdominal and low back/hamstring musculoskeletal function

As with the original test, the 1980 Health-Related Physical Fitness Test was norm-referenced. A *norm* is a statistic that describes how a large number of individuals of differing genders, ages, and abilities perform on a specific test. It allows students to compare themselves with a peer group.

In 1988 AAHPERD instituted Physical Best, a comprehensive physical fitness and assessment program designed not only to measure current fitness levels of children, but also to motivate them to participate in physical activity to become their *physical best*. Further, it is designed to raise student awareness of fitness and help them develop a sense of responsibility for their future health and quality of life. To date, Physical Best has been adopted by hundreds of schools and organizations throughout the United States. The complete program contains a health-related fitness assessment, an educational component, and an awards system.

The educational component contains a kit that is designed to help the physical educator:

- Learn how to motivate students to develop good fitness habits
- Design specific activities, both in and out of school, to help students improve
- Teach students to set individualized fitness goals based on their current fitness levels and their unique capabilities for improvement
- Monitor students' progress and keep fitness records
- Share results with parents through report cards and letters

The kit contains an instructor's manual with guidelines for developing effective fitness education and information needed to plan and implement the program. Computer software is available to aid in the storage and reporting of fitness data. Physical Best does not contain normative data but rather is a criterion-referenced test. Therefore, student performance is compared not to a peer group but rather to a preestablished criterion (score).

In 1994 AAHPERD formed a partnership with the Prudential FITNESSGRAM. This comprehensive fitness program consists of (1) a health-related fitness assessment; (2) a computerized reporting program; (3) a behavioral-oriented recognition system; and (4) educational materials. AAHPERD has maintained the Physical Best educational component and awards system but has replaced its assessment system with that contained in the FITNESSGRAM.

The Psychomotor Domain

Assessment in this domain deals with the acquisition of skills and can be accomplished both objectively and subjectively. *Objective assessment* involves the measurement of the *product*—how many baskets the student made, how fast the student completed the dribble maze, and how far the student threw the ball. It yields a *quantitative* score. *Subjective assessment* involves the measurement of the *process*—how well the student performed the skill while shooting, dribbling, and throwing. It yields a *qualitative* score. Suggested sources for a variety of skills tests are shown in table 11.3.

Evaluation of students in the elementary program (grades K–6) should be more process oriented. However, standardized skills tests can still be used with younger students if the tests are modified to meet their characteristics. Process assessment is done primarily by way of observing students as they perform the skills. To make the observation somewhat objective, a checklist or rating scale may be used, such as the examples shown in figures 11.2 and 11.3.

The evaluation of student performance and progress is an essential part of all physical education programs.

Table 11.3
Skills Tests Sources

Source
American Alliance for Health, Physical Education, Recreation, and Dance. Reston, VA: AAHPERD.
Baumgartner, T., Jackson, A., Mahar, M., & Rowe, D. (2003). *Measurement for evaluation in physical education and exercise science.* New York: McGraw-Hill.
Lacy, A., & Hastad, D. (2003). *Measurement and evaluation in physical education and exercise science.* San Francisco: Benjamin Cummings.
Miller, D. (2002). *Measurement by the physical educator: Why and how.* Dubuque, IA: William C. Brown.
Morrow, J., Jackson, A., Disch, J., & Mood, D. (2005). *Measurement and evaluation in human performance.* Champaign, IL: Human Kinetics.
Tritschler, K. (2000). *Barrow & McGee's practical measurement and assessment.* Philadelphia: Lippincott, Williams, & Wilkins.

Figure 11.2
Checklist for Overhand Throw

Name: _____ Grade: _____ Date: _____

Directions: While viewing the student perform the overhand throw, place a check (✔) in the appropriate space.

1. Steps with the proper foot   Yes ___   No ___
2. Leads with the elbow   Yes ___   No ___
3. Rotates at the trunk   Yes ___   No ___
4. Has complete follow through   Yes ___   No ___
5. Has smooth, continuous movement   Yes ___   No ___

Figure 11.3
Rating Scale for Overhand Throw

Name: _____ Grade: _____ Date: _____

Total rating:
1 = Does only one of five steps
2 = Does only two of five steps
3 = Does four of five steps
4 = Has a smooth, continuous, mature movement, doing all five steps

Directions: After viewing the student perform the overhand throw two or more times, circle the appropriate rating.

Action	Scoring
1. Ball held with the fingers	1 2 3 4
2. Steps with the opposite foot	1 2 3 4
3. Rotates at the hip	1 2 3 4
4. Leads with the elbow	1 2 3 4
5. Ends with a follow through	1 2 3 4

The Cognitive Domain

Assessment in this domain deals with the development and application of knowledge; that is, the history, rules, terminology, techniques, and strategy of selected activities; fitness and wellness concepts; and safety concerns. Knowledge assessment is just as important as the assessment of both fitness and skills, but too often it is ignored. It can be accomplished either formally or informally. Formal assessment involves the use of a teacher-created paper-and-pencil test and is most appropriate for grades 4–12, when students have developed reading and writing skills. When constructing a formal test, it is important to meet the criteria of validity, reliability, objectivity, and administrative feasibility.

For grades K–3, informal assessment in the form of oral questioning is recommended. Questioning students about the various concepts being taught allows the teacher to determine whether they are learning and understanding. This may be done as a review at the end of each day's lesson or during the lesson, when appropriate.

The Affective Domain

Assessment in this domain addresses the development of appropriate attitudes, values, and social behaviors. It involves the interests, appreciation, and emotional biases of a student. More specifically, the affective domain focuses on the development of sportsmanship, cooperation, teamwork, self-restraint, and leadership.

Miller (2002) states that many physical educators develop affective objectives for students, yet sometimes feel uncomfortable measuring students and therefore make no attempt to determine if these affective objectives have been attained. In addition, many physical educators think that if the cognitive and psychomotor objectives are reached, the affective objectives will somehow also have been accomplished. Even though the affective domain is the most difficult to measure objectively, Schiemer (2000) indicates there are observable behaviors that provide insight into a student's affective development, such as whether the student:

- Assists classmates
- Demonstrates self-control
- Displays sportsmanship
- Encourages and compliments teammates
- Follows directions
- Plays by the rules
- Shows respect for teacher and classmates
- Works cooperatively with a partner

Even though affective assessment is challenging, an attempt should be made to determine the feelings and appreciation levels that students take with them after exposure to physical education. Schiemer (2000) states that when assessing affective behaviors, it is important for the physical educator to provide appropriate movement experiences so students can learn and practice these behaviors. The following strategy is recommended:

- Identify an observable affective behavior
- Provide examples of both acceptable and unacceptable ways this behavior is displayed
- Have students practice the acceptable behavior
- Observe the behavior during a lesson

Affective development can be measured by (1) teacher observation of student behavior, (2) self-appraisal, or (3) a combination of the two. An example of an assessment tool for affective behavior by way of teacher observation is shown in figure 11.4. A modified version of a self-appraisal form developed by Schiemer (2000) is shown in figure 11.5. The author suggests that when this form is to be used, the physical educator provide activities that allow ample opportunities for cooperation. Examples of teacher-student appraisals are shown in figures 11.6 and 11.7.

There are physical education professionals who question whether teachers should undertake affective evaluation. They contend that physical educators do not have enough time to devote to this domain and that it is more important to teach motor skills. In addition, the availability and reliability of appropriate measuring instruments are questionable. Furthermore, Miller (2002) asks why physical education should assume responsi-

---

Figure 11.4
Student Behavior Checklist

Name: _____ Grade: _____ Date: _____

Check one:

____ Midterm assessment

____ Final assessment

Directions: After viewing the student's behavior, place a check (✔) in the appropriate space.

1. Assists classmates   Yes ____ No ____
2. Arrives promptly for class   Yes ____ No ____
3. Follows directions   Yes ____ No ____
4. Accepts criticism   Yes ____ No ____
5. Demonstrates self-control   Yes ____ No ____
6. Cooperates with others   Yes ____ No ____
7. Shows respect for others   Yes ____ No ____
8. Obeys class rules   Yes ____ No ____
9. Works hard to complete each task   Yes ____ No ____
10. Works independently   Yes ____ No ____

Figure 11.5
Cooperation Self-appraisal

Name: _____

Grade: _____

Elements of Cooperation:

- Follow rules
- Obey the teacher
- Help classmate(s)
- Encourage classmate(s)
- Compliment classmate(s)
- Shows concern for classmate(s)

Date: _____

Lesson: _____

Rate yourself on your cooperative behavior today.

Lowest   Highest
  1   2   3   4   5

Explain why you gave yourself this rating.

_____

_____

_____

_____

Source: Adapted from Schiemer, 2000.

Figure 11.6
Teacher-Student Affective Assessment

Name: _____

Grade: _____

2 = Exceeds Expected Behavior   1 = Meets Expected Behavior   0 = Needs Improvement

Behavior	1st 9 weeks		2nd 9 weeks	
	Student	Teacher	Student	Teacher
Actively participates				
Cooperates with classmates				
Is respectful to teacher				
Follows directions				
Demonstrates self-control				
Displays sportsmanship				
Works independently				

Figure 11.7
Teacher-Student Affective Domain Assessment

Name: _____

____ Self assessment

____ Teacher assessment

Assessment:

Date #1: _____ Date #2: _____

Criteria:	Rating #1	Rating #2
Etiquette		
• Respects others' personal space		
• Honors activity dynamics		
• Conforms to standards of conduct		
Fairness		
• Plays fair		
• Accepts defeat and does not complain		
• Accepts victory and does not gloat		
Communication with Peers		
• Encourages others		
• Accepts difficult individual skill levels		
• Assists others in reaching success		
• Uses positive words and body language respectfully		
Communication with Teacher		
• Uses positive words and body language respectfully		
• Accepts teaching cues in a positive manner		
• Responds to instruction		
• Remains on task		

Source: A. Gallo. 2003. Assessing the affective domain. *Journal of Physical Education, Recreation and Dance* 74(4): 44–48. Reprinted with permission of JOPERD.

bility for developing and assessing affective behavior when other disciplines like English, mathematics, and the sciences rarely do. Others argue just as persuasively that affective learning and behavior should be evaluated. The affective development that results from varied experiences in a physical education program is important. The feelings and appreciation derived from physical education participation shape a student's attitudes and these attitudes, in part, determine if the student continues to be physically active after leaving school.

EVALUATION OF STUDENTS WITH DISABILITIES

As previously stated, student evaluation is an essential responsibility of today's teachers. Perhaps nowhere in the curriculum is this more important than in the adapted physical education program. Assessment in the adapted program serves the following purposes:

1. *Screening*—To determine which students are in need of special help
2. *Placement*—To assure that each student needing special help is in the proper environment
3. *Diagnosis*—To determine the present level of performance of each student to guide both the selection of activities and instruction
4. *Progress*—To determine if the behavioral objectives have been met

Existing tests in the four domains should be used for assessing students with disabilities. However, these tests may not always be appropriate. A list of the tests most frequently used by physical educators for the screening, placement, diagnosis, and progress assessment of students with disabilities is given in table 11.4.

Table 11.4
Assessment Tests for Students with Disabilities

AAHPERD Motor Fitness Testing Manual for the Moderately Mentally Retarded (Johnson & Londeree, 1976)

Andover Perceptual-Motor Test (Nichols, Arsenault, & Giuffre, 1980)

Ayres Southern California Perceptual-Motor Tests (Miller & Sullivan, 1982)

Basic Motor Ability Tests (Arnheim & Sinclair, 1979)

Brigance Diagnostic Inventory of Early Development (Brigance, 1978)

Bruininks-Oseretsky Test of Motor Proficiency (Bruininks, 1978)

Denver Developmental Screening Test (Frankenburg & Dodds, 1967)

Fait Physical Fitness Test for Mildly and Moderately Mentally Retarded Students (Dunn & Fait, 1989)

Hughes Basic Gross Motor Assessment (Hughes, 1979)

I CAN Fundamental Skills Test (Wessel, 1979)

Kansas Adapted/Special Physical Education Test Manual (Johnson & Lavay, 1988)

Ohio State University Scale of Intra-Gross Motor Assessment (Loovis & Ersing, 1979)

President's Challenge for Students with Special Needs (President's Council on Physical Fitness and Sports, 1991)

Project ACTIVE Motor Ability Test (Vodola, 1976)

Projective Unique Physical Fitness Test (Winnick & Short, 1985)

Prudential FITNESSGRAM Modifications for Special Populations (Cooper Institute for Aerobics Research, 1992)

Purdue Perceptual-Motor Survey (Roach & Kephart, 1966)

Special Fitness Test Manual for Mildly Mentally Retarded (AAHPERD, 1976)

Test of Gross Motor Development (Ulrich, 1986)

PORTFOLIO ASSESSMENT

A current attempt to provide a broader, more genuine view of what a student has learned is called *portfolio assessment* (Zessoules & Garner, 1991). It is a move away from the more traditional evaluation process of cognitive-knowledge test, psychomotor-skills test, or affective-teacher rating scale.

Portfolio assessment is a process for documenting what Perrone (1991) calls *authentic learning*—that is, learning that occurs in the natural setting. Smith (1997) stipulates that as a collection of student work, the portfolio should incorporate several qualities so as to truly reflect the student's ability, capacity, and/or achievements. Fulmer (1994) adds that it should cover the breadth and scope of the student's curricular experiences. This alternative approach stems from the notion that assessment should relate directly to the outcomes deemed important and relevant. Siedentop (2004) provides the following example. In assessing how well a student can write, rather than have the student complete a multiple-choice exam on writing, the authentic assessment movement requires the student to write an essay. In physical education, rather than record the number of tennis strokes a student can successfully perform against a rebound board in 30 seconds, the student is judged on the forehand and backhand techniques during game play. While skills testing is not inappropriate for a portfolio, a variety of other assessment items also may be used to make up a portfolio:

- Activity diaries
- Attitude inventories
- Entry-level skills tests
- In-class assignments
- Independent study contracts
- Knowledge tests
- Parental questionnaires
- Quizzes
- Research papers
- Self-assessment checklists
- Student reflection reports
- Task sheets
- Teacher/peer rating forms
- Unit logs

The specific assessment items selected for a portfolio may be determined by school policy, but Defina (1992) argues that this determination should be made, in part, by the student. Allowing students to make decisions on what they want to learn and how they will know if they have learned it is an important component of portfolio assessment.

Melograno (1994) indicates that portfolios provide a nonstigmatizing, motivational, effective means for reporting student learning. When used correctly, portfolios can:

- Capitalize on student work
- Enhance both teacher and student involvement in evaluation
- Satisfy the accountability need prompted by school reform (Chittenden, 1991)

TEACHER EVALUATION

The teaching profession is scrutinized more closely today than ever before. With increased emphasis on teacher accountability, there is a corresponding effort to evaluate those responsible for educating our youth.

Barrow, McGee, and Tritschler (1989) stress that in order to improve the program, educators need to focus on improving the instructional process. Teacher evaluation is the best means of determining the value of the instructional process in physical education. Beyond this purpose, Castetter (1971) lists the following reasons for the ongoing evaluation of teachers:

- To improve instruction
- To determine teacher potential
- To determine teacher strengths and weaknesses
- To promote self-development
- For the assignment of teacher responsibilities
- To determine salary increases
- For retention determination

Perhaps the first step in a meaningful teacher evaluation is the development of a valid assessment procedure. The more common, traditional procedures for teacher evaluation are (1) self-appraisal, (2) administrator checklist/rating, (3) systematic observation, and (4) student ratings. Self-appraisal requires the teacher to analyze his/her teaching behavior. Thomas (1980) indicates that self-appraisal is necessary for improving instruction and that a simple reflection about course objectives, student interest, and student participation can offer useful insights for teaching improvement. An example of a teacher self-appraisal form is shown in figure 11.8.

Administrator checklist/rating is, perhaps, the most frequently used procedure for evaluating teachers. A checklist or rating form contains a group of statements about which the administrator gives his/her opinion relating to the teaching-learning environment. When using a checklist, the judgment usually is in the form of yes or no answers. With a rating form, the range of responses is greater. An example of an administrative rating form is shown in figure 11.9.

Student ratings of teacher effectiveness are widely used, especially in higher education. Their use should be guarded because of the extraneous variables that may affect how a student responds to the selected components of teaching. Among these variables are class size, class composition (coeducational/noncoeducational), and whether the class is required or an elective. There are those who caution against the use of student ratings because (1) regardless of age, they feel students are not mature enough to make valid judgments regarding the quality of instruction; and (2) if students are used for assessing teaching behavior, a teacher may seek popular-

Figure 11.8
Teacher Self-Evaluation Checklist

Name: _____

Date: _____

Check one:

_____ End of first grading period

_____ Mid-semester

_____ End of semester

Directions: Answer each question with either Yes or No. Five points are allotted for each Yes. A score of sixty (twelve of fifteen) is needed for an above-average teacher.

1. Do I know the names of all my students?
2. Do I get my students to think instead of giving back rote memory of what I say?
3. Do I analyze my own classroom or gymnasium teaching?
4. Do I find myself available for student conferences?
5. Do I listen attentively to what students are saying?
6. Am I intuitive to the varying needs of students, such as hypertension?
7. Do I know my subject matter well enough to be challenged by questions in class?
8. Do I really try to develop a positive teaching-learning climate in my classes?
9. Do I give individual attention to the excellent, the moderate, and the slow learner, and not concentrate just on the talented ones?
10. Do students talk to me freely inside and outside of class?
11. Are my students sleeping in class or getting restless?
12. Am I tolerant of students' mistakes as well as my own?
13. Do I keep up to date with my professional reading?
14. Do I grade fairly on learning objectives rather than on likes and dislikes?
15. Do I really care about students and let them know it?

ity at the expense of effective teaching. An example of a student rating form is shown in figure 11.10.

Systematic observation has been popular in recent years. It requires both observation and documentation by another individual, usually an administrator or a colleague. To promote its use, Darst, Zakrajsek, and Mancini (1989) state that systematic observation allows an individual, while following stated guidelines and procedures, to observe, record, and analyze interactions with the assurance that others viewing the same events would agree with this recorded data. A list of instruments for systematic observation includes the following:

- Academic Learning Time-Physical Education (ALT-PE)
- Cheffer's Adaptation of the Flanders's Interaction Analysis System (CAFIAS)
- Flanders's Interaction Analysis System (FIAS)

Figure 11.9
Administrative Rating Form of Teacher Behavior

Directions: While or after viewing the educator in a teaching setting, score each of the items below accordingly.

Scoring:
5 = Superior
4 = Acceptable
3 = Needs improvement
2 = Unsatisfactory
1 = No opportunity to observe

Characteristic	Score	Characteristic	Score
Personality	_____	Knowledge of subject matter	_____
Appearance	_____	Ethical behavior	_____
Health	_____	Punctual and dependable	_____
Enthusiasm	_____	Teaching	_____
Initiative	_____	Planning and organization	_____
Flexibility	_____	Voice level and modulation	_____
Poise and self-confidence	_____	Nonverbal communication	_____
Professionalism	_____	Motivation of students	_____
Ability to work with others	_____	Attention to individual needs	_____
Understanding of youth	_____	Discipline	_____

- Observational Recording Record of Physical Educator's Teaching Behavior (ORRPETB)
- Ohio State University Teacher Behavior Rating Scale (OSUTBRS)
- Rankin Interaction Analysis System (RIAS)

Regardless of the technique used, Bucher and Koenig (1983) provide the following guidelines for teacher evaluation. The evaluation should:

- Involve the teacher personally. Since evaluation is a cooperative venture, the teacher should not only understand the process but should also be involved in developing the evaluative criteria.
- Focus on teacher performance
- Be concerned with helping the teacher grow professionally
- Look to the future and be concerned with improvement of the program
- Be well organized and administered. The procedure should be clearly outlined with a step-by-step approach

PROGRAM EVALUATION

The purpose of this book has been to examine the total physical education curriculum. Although its parts have been discussed, the overall objective has been to relate specific aspects to the entire program and to effect a

Figure 11.10
Student Rating Form for Teacher

Name: _____

Course: _____

Date: _____

Directions: This evaluation is made at the instructor's request for determining teaching effectiveness. Listed below are qualities that describe aspects of instructor performance. Please rate the instructor on each of these items by circling your honest response.

Scoring:

5 = Excellent
4 = Very good
3 = Good
2 = Fair
1 = Poor

The instructor:

1. Clearly presented course objectives.	5 4 3 2 1
2. Used appropriate evaluative procedures.	5 4 3 2 1
3. Stimulated my interest in the subject matter.	5 4 3 2 1
4. Used resource materials well.	5 4 3 2 1
5. Broadened my interest in the subject matter.	5 4 3 2 1
6. Displayed enthusiasm for teaching.	5 4 3 2 1
7. Communicated effectively.	5 4 3 2 1
8. Made appropriate examples of content.	5 4 3 2 1
9. Taught to the level of skill and/or knowledge of the class.	5 4 3 2 1
10. Was willing to give advice and attention according to my individual needs.	5 4 3 2 1
11. Inspired confidence in me relating to course content and its application.	5 4 3 2 1
12. Motivated me to apply the subject matter to my own life.	5 4 3 2 1
13. Required too much work for the credit.	5 4 3 2 1
14. Used lecture and discussion well.	5 4 3 2 1
15. Is recommended to my friends.	5 4 3 2 1

proper balance between grade level progression, developmental and adapted class instruction, intramurals, and interscholastic athletics. No one area has been singled out as the most important because they are all important.

The primary purpose of program evaluation is to achieve program effectiveness. School administrators and directors of physical education want to know how well their programs are working and how they compare with others in neighboring communities, in neighboring states, or on a national level. Program effectiveness may be determined by a self-appraisal checklist or inventory or student performance.

Using a self-appraisal checklist or inventory allows individuals to compare the component parts of the curriculum to recognized, acceptable standards. Its focus is on the physical characteristics of the program, that is to say whether the environment (curriculum, equipment, facilities, faculty, etc.) meets specified criteria. Baumgartner, Jackson, Mahar, and Rowe (2003) indicate that the environment is one of the least important variables in evaluating a program. Still, self-examination is often a good practice since it encourages soul searching and an honest evaluation of what the program should be. When administrators or faculty members appraise the total curriculum, they have opened their minds and humbled themselves just enough to be able to see the changes needed. Therefore, the simple act of deciding to appraise a program is significant. It is the first step to program modifications in a changing society.

There are a number of self-appraisal instruments designed for the physical education program; several are standardized and have had widespread use. State Departments of Education in California, Florida, Idaho, Illinois, Indiana, Kansas, Ohio, Texas, and Wisconsin have taken the lead in developing instruments for this kind of appraisal.

The *LaPorte Health and Physical Education Score Card I, II* (1951) has been in use longer than any other similar instrument. It is intended to measure all aspects of the health, safety, recreation, and physical education program. Attention is focused on the characteristics of a good program. Ten program elements are identified and can be rated from 1–30, with a perfect score totaling 300. If 200 points are scored, the program is considered fair to good. If 100 points are scored, the program is considered poor.

Bucher (1977) developed the lengthy but complete *Checklist and Rating Scale for the Evaluation of the Physical Education Program*. It combines the standards required by several state departments of education with standards recommended by leading authorities in the field. This instrument is properly suited for the appraisal of a total school program. Not all of Bucher's scale needs to be used for any particular school. Ample items exist in each of the following sections to enable a school to appraise individual components of its program:

- *General administrative considerations*—including philosophy, objectives, and the general curriculum
- *Specific administrative considerations of the physical education program*—including attendance, excuses, required classes per week, length of class, time provisions for showering and dressing, class size, student-to-teacher ratio, public relations, athletic standards, supervision of intramural and extramural activities, teacher scheduling, coeducational instruction, and supervision
- *Components of the physical education program*—including program design and activities offered in the instructional, intramural, and interscholastic athletic programs
- *Staff*—including status, tenure, and background preparation

- *Facilities and equipment*—including general standards, equipment, indoor and outdoor facilities, faculty offices, locker rooms, and showers
- *Measurement and evaluation techniques*—including student status and progress as well as staff evaluation

AAHPERD (1977) published the *Assessment Guide for Secondary School Physical Education Programs*. Its purpose is to provide a functional, easily administered, comprehensive tool that allows for:

- Self-study and self-evaluation
- Identification of problem areas
- Improvement of the program

The instrument is composed of statements relating to administration, the athletic program, the instructional program, and the intramural program. A *yes* or *no* response is required for each of the 46 statements. A negative response for an item may indicate that additional self-study is needed or may be justified with an acceptable rationale based on the philosophy of the individual department.

In 1994 AAHPERD published the *Checklist for Elementary School Physical Education*, developed by NASPE's Council on Physical Education for Children. It is comprised of 94 items that are divided into the following seven categories:

- Teacher
- Teacher preparation/staff development
- The instructional program
- Assessment
- Organization/administration
- Equipment and facilities
- School-related programs

Evaluative criteria are presented in the form of statements describing desirable attributes. The evaluator is supposed to determine the extent to which each statement describes the program in question. The eighteen items dealing with the instructional program are shown in figure 11.11.

Regardless of the technique or instrument used to assess the program, the collected data should allow an individual to determine (1) the level of student achievement, (2) the strengths and weaknesses of the program objectives, (3) the quality of instruction, and (4) the extent of facility and equipment utilization. In addition, it is recommended that if self-appraisal is used, school administrators and the entire physical education staff should participate in the assessment.

Assessing student performance to determine program effectiveness as a measure of accountability is the most valid index of the quality of a program. The use of a student's data in the areas of fitness, skill, knowledge, and attitude allows for curriculum revision to better attain stated objectives, allows for identification and elimination of undesirable outcomes, and serves as an effective means for both the administration and the com-

Figure 11.11
The Instructional Program

Read each statement carefully and circle the letter that is appropriate.

A—If the attribute is in consistent agreement with what is present in the physical education program.

B—If the attribute is in general agreement but the need may exist for improvement in this area.

C—If you are undecided as to whether or not the item describes the physical education program.

D—If you definitely disagree that the attribute describes what is actually present in the physical education program and a definite need exists for improvement in this area.

E—If no basis for evaluation exists.

The physical education program:

1. Is an integral part of, and consistent with, the total educational philosophy of the school.   A B C D E
2. Serves the diverse needs of all students   A B C D E
3. Is based on an established written curriculum.   A B C D E
4. Is of sufficient breadth and depth to be challenging to all.   A B C D E
5. Is developmentally based and progressively sequenced from year to year.   A B C D E
6. Is regularly updated and revised.   A B C D E
7. Has well defined objectives for progressive learning.   A B C D E
8. Is built around the development of efficient, effective, and expressive movement abilities.   A B C D E
9. Provides opportunities for the development of fundamental movement patterns and specific movement skills.   A B C D E
10. Provides experiences to enhance the development of physical fitness.   A B C D E
11. Provides opportunities for students to develop skills in games/sports, dance/rhythms, and gymnastics.   A B C D E
12. Provides opportunities for students to enhance their knowledge and understanding of human movement in a variety of physical activities.   A B C D E
13. Promotes safe behavior by incorporating proper safety practices into physical education lessons.   A B C D E
14. Fosters creativity.   A B C D E
15. Promotes self-understanding and acceptance.   A B C D E
16. Promotes positive social interaction and self-control.   A B C D E
17. Recognizes and provides for learning enjoyment and fun.   A B C D E
18. Helps each student learn how to manage risk-taking and other challenges.   A B C D E

From NASPE's *Checklist for elementary school physical education*; available from AAHPERD by calling 1-800-321-0789. Reprinted by permission.

munity to monitor schools. In general, student data allows for quality control, which is important to any educational program. Quality control is a way of determining whether the program is well executed and whether it remains as effective over time as it was when first implemented.

SUMMARY

1. Evaluation is the means by which the physical educator demonstrates accountability.

2. Evaluation is a three-step process involving (1) data collection (measurement), (2) judging the value of the data, and (3) making a decision.

3. Beyond determining the status, progress, and/or achievement of students, evaluation serves other purposes that include classification, prediction, and motivation.

4. Evaluation in the school setting involves the student, the teacher, and the program.

5. Student evaluation occurs in four areas: physical, psychomotor, cognitive, and affective.

6. Teacher evaluation, primarily designed to improve instruction, has been traditionally accomplished by self-appraisal, professional rating, and/or student ratings. Systematic observation has gained recent popularity.

7. Program evaluation, designed to determine program effectiveness, may be accomplished by student performance and self-appraisal.

QUESTIONS AND LEARNING ACTIVITIES

1. Visit a local industry and inquire about quality control practices. How do these efforts compare with those employed in public schools?

2. Physical education philosophers have written about the place and importance of measurement in the program. What seems to be the prevailing viewpoint today?

3. How do you feel about establishing outcomes for boys and girls in grades K–12? Do students readily accept that these outcomes are worthy of in-class and out-of-class activity?

4. At what point in the evaluation process do measurement activities become too time consuming? Explain your viewpoint and illustrate it with an example.

5. Explain how you may use the results of measurement in a secondary school to revise a given curriculum.

6. Examine curriculum guides from various states and/or from other communities. Do you find that measurement and evaluation are

considered part of the total program? How do your findings compare with those of your classmates?

7. Obtain a copy of one of the program evaluation instruments referred to in this chapter. Use this instrument to critically evaluate an existing program. Provide a list of strengths and weaknesses of the program, as well as recommendations to improve the program.

8. Develop a short statement relating to the significant relationship between educational aims and objectives and evaluation practices. Before doing this, read what various authors have said about this topic.

9. Select an imaginary school system in a city of 50,000 people. Assume you are going to revise the curriculum partly on the basis of measurement findings. What procedures would you follow? Illustrate a plan for relating evaluation results to curriculum revision.

10. Prepare a proposed plan for grading a seventh grade physical education class. Include the components (skill, knowledge, and so on) and how each will be measured.

REFERENCES

American Alliance for Health, Physical Education, Recreation, and Dance. (1976a). *Special fitness test manual for the mildly mentally retarded*, 2nd ed. Reston, VA: Author.

American Alliance for Health, Physical Education, Recreation, and Dance. (1976b). *Youth fitness test manual*. Reston, VA: Author.

American Alliance for Health, Physical Education, Recreation, and Dance. (1977). *Assessment guide for secondary school physical education programs*. Reston, VA: Author.

American Alliance for Health, Physical Education, Recreation, and Dance. (1980). *Health-related physical fitness test manual*. Reston, VA: Author.

American Alliance for Health, Physical Education, Recreation, and Dance. (1987). *Program appraisal checklist for elementary school physical education programs*. Reston, VA: Author.

American Alliance for Health, Physical Education, Recreation, and Dance. (1988). *Physical best*. Reston, VA: Author.

Barrow, H., McGee, R., & Tritschler, K. (1989). *Practical measurement in physical education and sport*. Philadelphia: Lea & Febiger.

Baumgartner, T., Jackson, A., Mahar, M., & Rowe, D. (2003). *Measurement for evaluation in physical education and exercise science*. New York: McGraw-Hill.

Bucher, C. (1977). *Administration of school and college health and physical education programs*. St. Louis, MO: Mosby.

Bucher, C., & Koenig, C. (1983). *Methods and materials for secondary school physical education*. St. Louis, MO: Mosby.

Castetter, W. (1971). *The personnel function in public administration*. New York: Macmillan.

Chittenden, E. (1991). Authentic assessment, evaluation, and documentation of student performance. In V. Perrone (Ed.), *Expanding student assessment*. Alexandria, VA: Association for Supervision and Curriculum Development.

Darst, P., Zakrajsek, D., & Mancini, V. (1989). *Analyzing physical education and sport instruction*. Champaign, IL: Human Kinetics.

Defina, A. (1992). *Portfolio assessment: Getting started*. New York: Scholastic.

Dunham, P. (1994). *Evaluation for physical education*. Englewood, CO: Morton.

Fulmer, R. (1994). *Secondary portfolios*. Ticonderoga, NY: Edlink.

Gallo, A. (2003). Assessing the affective domain. *Journal of Physical Education, Recreation, and Dance, 74*(4), 44–48.

Melograno, V. (1994). Portfolio assessment: Documenting authentic student learning. *Journal of Physical Education, Recreation, and Dance, 65*(8), 50–55, 58–61.

Miller, D. (2002). *Measurement by the physical educator: Why and how*. Dubuque, IA: William C. Brown.

Perrone, V. (Ed.). (1991). *Expanding student assessment*. Alexandria, VA: Association for Supervision and Curriculum Development.

Schiemer, S. (2000). *Assessment strategies for elementary physical education*. Champaign, IL: Human Kinetics.

Siedentop, D. (2004). *Introduction to physical education, fitness, and sport*. New York: McGraw-Hill.

Smith, T. (1997). Authentic assessment: Using a portfolio card in physical education. *Journal of Physical Education, Recreation, and Dance, 68*(4), 46–52.

Thomas, N. (1980). Thoughts on teacher evaluation. *The Physical Educator, 37*(4), 176–178.

Zessoules, R., & Garner, H. (1991). Authentic assessment: Beyond the buzzword and into the classroom. In V. Perrone (Ed.), *Expanding student assessment*. Alexandria, VA: Association for Supervision and Curriculum Development.

INDEX

271